We

Peo

s

Lawrence, Kansas 66046

CMP Books
CMP Media, Inc.
1601 W. 23rd Street, Suite 200
Lawrence, KS 66046
USA
www.cmpbooks.com

Acquisitions Editor:	Berney Williams
Editor:	Liza Niav
Layout Production:	Kristi McAlister
Cover Art Design:	© FPG International LLC.

Distributed in the U.S. and Canada by:
Publishers Group West
1700 Fourth Street
Berkeley, CA 94710
1-800-788-3123
www.pgw.com

ISBN: 1-929629-08-7

R&D Developer Series

To my wife Kim,
whose selflessness is second to none.
To my children, Shane, Ryan, Lindsey, and Kyle —
the pride and victories of my life.

In memory of my good friend,
Yuri Rubinsky,
the silent advocate of accessibility for people
with disabilities.

Table of Contents

Part III Development Resources

Foreword

The social and economic impact of the Internet and Information Technology (IT) has made its presence clear. Over the past several years, information technology has accounted for one-third of U.S. economic growth. To compete and win in today's global marketplace, companies are using IT to develop products more rapidly, forge closer relationships with their customers and their suppliers, and deliver "just-in-time" training to their employees. One hundred and fifty-eight million people now have access to the global Internet — 79 million in the United States alone. Electronic commerce is beginning to explode, and it could exceed $1.3 trillion in the United States by 2003.

In addition to its commercial significance, IT can also be a powerful tool for addressing some of our most pressing social challenges. In some schools, children are using the Internet to take field trips to the bottom of the ocean, tap into the Library of Congress to research a history paper, or use a remote telescope to explore the universe. People in rural communities can access better quality health care through telemedicine. Governments can be more open, responsive, and provide better services to their citizens — since there is no line online.

The incredible economic and social potential of IT makes it all the more important that it be accessible for people with disabilities. If it is accessible, IT could significantly improve the quality of life for people with disabilities by increasing their independence and their ability to participate in the workforce. If it is not, it could further isolate people with disabilities, and prevent them from being full participants in the Information Society.

That's why President Clinton and Vice President Gore have pushed to ensure that the World Wide Web and other computing and communications technologies are accessible. The Web Accessibility Initiative (WAI) of the World Wide Web Consortium grew out of a meeting held at the White House. President Clinton strongly endorsed WAI in 1997, stating that "it is vital that the Web be accessible to everyone." U.S. government agencies (the National Institute on Disability and Rehabilitation Research and the National Science Foundation) joined with leading IT companies and the European Commission to fund this important initiative.

On January 13, 1999, President Clinton unveiled a major initiative to increase employment opportunities for people with disabilities. As part of this effort, he announced new steps to ensure that IT is accessible to people with disabilities — including making the federal government a model user of accessible technology, providing low-interest loans to people with disabilities who might not otherwise be able to afford accessible technology, and investing in research and development to improve state-of-the-art accessible technology such as speech and gesture recognition, text-to-speech, and automatic captioning.

Making the Web accessible for everyone is critical. It will help not only the hundreds of millions of people worldwide with disabilities, but also mobile professionals who may be trying to access the Web in a "hands-busy, eyes-busy" environment or by using a cellular phone. Making the Web accessible is not only the ethical thing to do, but in many countries, it's the law.

That's why Mike Paciello's book is so important. We can't have an accessible Web unless everyone who is building the Web understands the importance of making it accessible, and has the technical know-how to do it. I hope that everyone who is building the Web will read this book and take its lessons to heart.

Thomas Kalil
Special Assistant for Economic Policy to President Clinton

Preface

The World Wide Web is the quintessential information broker, the ultimate library, the metacenter of information technology. It is the epitome of a "give free, receive free" model where anyone can create content for the Web and anyone can retrieve that same information. Better still, the Web is a virtual world of communities, businesses, schools, and governments all seamlessly connected and — at first glance — completely accessible to every human.

However, the Web is really *not* accessible to every human. Rather, the Web is a living dichotomy: It is a highly interactive, advanced technology designed to increase communication and collaboration. But the user interface is complex, lacks intuitiveness, and is certainly not easy to use by many human communities — particularly those with disabilities.

Overlaid with a graphical user interface (GUI), blind users find the Web increasingly difficult to access, navigate, and interpret. People who are deaf and hard of hearing are served Web content that includes audio but does not contain captioning or text transcripts. Web devices and appliances (for example, WebTV and Web kiosks) present challenges to people with physical disabilities or repetitive strain injuries (RSI). These Web devices are often hard to access by a person in a wheelchair and contain controls, buttons, and touch screens that are difficult to manipulate.

The future indicates more challenging user requirements. For example, new desktops that implement a natural speech interface. Designed to remove the mundane chore of the "point and click" mouse action, this emerging technology could make it next to impossible for the deaf and people with speaking disabilities to communicate on the Web.

If the Web existed simply as an advanced information technology shared and used primarily by "geeks" or "Webheads," accessibility would be of little importance. But the Web is emerging as a critical path to practically every aspect of human life. Banking, consumer product sales, federal aid, healthcare, and entertainment are just a few of the services that have moved toward the one-to-one customer paradigm shift. The result is direct customer interaction that is efficient, expeditious, cost-effective and, most of all, profitable.

The fact is that the Web is not as accessible as many would have you believe. These critical citizen and customer services are not easily available to an individual with a disability. Worse yet, the geeks and Webheads responsible for the next generation Web have little or no idea how to make the Web accessible to the disabled. To further complicate the situation, people with disabilities, government agencies, and civil rights activists are beginning to use the legal system to force accessibility . . . and they are winning.

Who Are You?

Web Accessibility for People with Disabilities is a comprehensive resource for Internet and Web administrators and developers who are faced with the challenge of providing access to their site for people with disabilities. This book's goal is to aid and encourage you through the process of making your site accessible to people with disabilities. To accomplish this, I guide you through the design and setup process, describing commonly available tools and utilities that have been specifically created to implement assistive aids (several of which are available through WebABLE!). Additional material including information, development, research, and product solution providers is also included.

What You'll Learn

Web Accessibility for People with Disabilities provides a unique perspective of the Information Superhighway. A first of its kind, this book describes the world and challenges of people with disabilities and accessibility to the World Wide Web. The intent of the book is to serve as a vital resource to people *without* disabilities. In effect, the book "welcomes you to our world."

WebABLE! focuses on the issues that employers, Internet service providers, and Web administrators face when dealing with Internet and Web access for people with disabilities. It provides insight on the legal parameters involving Web accessibility and why you may be required by law to ensure access to your Web site.

This book provides answers to questions like: How can I provide highly graphic and visual content to the blind? How will people with physical disabilities access Internet kiosks? What will the deaf do when faced with voice output and audio-based multimedia they cannot hear? How will individuals who cannot speak interact with an information kiosk that is built on a voice-recognition platform?

This book relies heavily on the Web for its resource material. Most standards, laws, references, guidelines, hints, and tips can be found on the Web. All of this is due to the hard work of many throughout the world who have shared these invaluable resources on their Web sites.

Lastly, this book promotes Web accessibility for all people. In spite of its inherent barriers, people with disabilities make up a large segment of the Web user population. Why? Because the Web clearly offers *everyone* an opportunity to enhance his or her life. An accessible Web opens new doors to communication, education, employment, and entrepreneurial opportunities — regardless of ability. On the accessible Web, there are people with different abilities — not disabilities. You have no idea that the person you are interacting with is blind, deaf, or physically challenged. Chances are, you don't care either. That's the beauty of the Web, particularly to a person with a disability. The accessible Web breaks down the myths and misconceptions about a person's physical or mental ability. Perhaps that is the greatest barrier of all. This is the goal of the book — to make the Web free of all barriers. To make it WebABLE!

How This Book Is Organized

Part I examines the key reasons for building an accessible Web. I discuss the ethical, legal and technological motivations for designing accessibility, specifically addressing the user needs of people with disabilities. This section provides statistical information about the disabilities population as well as an in-depth description of the kind of barriers people with disabilities must overcome in order to use the World Wide Web as it exists today. Part I also emphasizes how businesses in the United States can be held liable by virtue of legal requirements, policies, and standards including the Americans with Disabilities Act (ADA) for not designing accessible Web sites. This section concludes with an in-depth discussion of the Web's barriers for people with disabilities.

Part II is a guide for building an accessible Web site. Web site administration involving specialized disability requirements, browsers, and content design is explained. Part II describes the various authoring tools, utilities, and validation services that Web site administrators can use to ensure the accessibility of their Web site.

Part III examines issues, concepts, standards, and products involving emerging technology and the World Wide Web, and concludes with a complete reference listing of various accessibility product solution providers, research centers, organizations, and general disability Web sites.

The Appendixes provide reference additional reference information, as well as a glossary of terms.

How to Use This Book

This book uses the following special symbols and features to highlight particular areas of interest.

Tip

This icon provides special hints, tricks, or insight to an accessibility feature. To help you develop good accessibility habits, I introduce specialized techniques and tips developed by the Web Accessibility Initiative (WAI) working groups.

Note

This icon identifies terms and concepts related to the world of accessibility, acronyms associated with the World Wide Web, and other noteworthy information that I hope you find interesting.

Cross-Reference

I've highlighted information that you can access — either on the Web or through other sources — with a cross-reference icon.

Contacting Me

Designing accessible Web sites is a work in progress. Advancements are made daily, almost as fast as the Web is being developed itself. Should you require updated information or general help in developing an accessible Web site, please feel free to contact me by sending e-mail to paciello@webable.com. You can also surf over to my Web site at http://www.webable.com.

Acknowledgments

This book has been a dream of mine for several years. Seeing it to completion could never have happened without the help and support of more people than I could possibly name in a few short lines.

Thanks to everyone in the disabilities and technology community who contributed to this book and to WebABLE. Many thanks to every Web site creator whose home page has contributed to the advancement of the accessibility of the Web and ultimately to this book's publication. All of the credit goes to you. And thanks to all the members of the WAI interest and working groups — the Web is a great place to live because of you!

Berney Williams and Paul Temme, the key folks at CMP who encouraged me from the start to write this book, found the way to make it happen, and then did all the right things to keep me plugging away at it. I am indebted to you and your colleagues at CMP whose enthusiasm, perseverance, and guidance were indispensable.

John Osborn, who started me on this journey and stuck with me when others didn't believe.

Ann Navarro and Kathy Gill were chiefly responsible for writing and editing Chapters 5, 8, and 9. Like John, I am indebted to you both.

Harvey Bingham, lead technical reviewer for this book and a colleague second to none.

Jim Miller, Tom Kalil, and Daniel Dardailler, who dedicated months of their time to work with me to create and launch the Web Accessibility Initiative, still the hallmark of my career.

Tim Berners-Lee, Jean-Francois Abramatic, Robert Cailliau, Sally Khudairi, Dave Raggett, Dan Connolly, Chris Lilley, Henrik Frystyk-Nielsen, Judy Brewer, and Susan Hardy — thank you for your enthusiastic support of the WAI.

Joseph Hardin, a good friend and staunch supporter of access for people with disabilities.

George Kerscher, T.V. Raman, Jim Allen, Gregg Vanderheiden, Murray Maloney, Molly Mannon, Charles Goldfarb, Harry Murphy, Joe Sullivan, Jon Gunderson, Mary Evans, Gary Moulton, Greg Lowney, Earl Johnson, Rick Brown, Ken Berk, Cynthia Waddell, Peter Sharpe, Tim Bray, Eve Maler, Bob Yeraska, Lee Fogal, Kevin St. George, Paul Hammerstrom, Herman Tavani, Dennis Wixon, Tom Spine, Mary Utt, Rick Frankosky, Betsy Comstock, Charles Abernethy, George Casaday, Marc Chardon, Sue Gault, Karen Shor, Minette Beabes, Jim Fruchterman, Jim Thatcher, Jason White, Hiroshi Kawamura, Josh Krieger, Danny Hilton-Chalfen, Norm Coombs, Dick Banks, Jim Rebman, Mark Novak, Jutta Treviranus, Jaron Lanier, Sandy Ressler, Larry Scadden, Frank Bowe, Fred Leung, Constantine Stephanidis, Michael Pieper, Clas Thorén, Jim Isaak, Curtis Chong, Deb Kaplan, Uli Strempel, Doug Wakefield, Judy Dixon, Dave Jaffe, Alan Cantor, Larry Goldberg, Jan Engelen, Tom Wesley, Wendy Chisolm, Dave Bolnick, and "Willie" Walker — just a handful of technologists I've had the great privilege of working with over the past 17 years.

Colin Moock for the original WebABLE! site design. Karin Trgovic for developing the original WebABLE! accessibility database. Kevin Nguyen and Kris Coward for the WebABLE! site administration. The University of Toronto and their Adaptive Technology Resource Center (ATRC).

Mac and Marti McCuller, who have hosted WebABLE for the past couple of years and supported my accessibility mission.

The "Friends" who supported the short-lived Yuri Rubinsky Insight Foundation. To all of you, everywhere. Special thoughts to Anna and Andrew (Yuri's parents) and Holley (Yuri's wife).

Yuri Rubinsky, who challenged the great technologists and visionaries to make this world accessible to all people.

My mom, the epitome of dedication and devotion.

Lastly, I want to thank my wife Kim and my children, Shane, Ryan, Lindsey, and Kyle.

About the Author

Michael Paciello is Chief Accessibility Officer of WebABLE, Inc. WebABLE is dedicated to stimulating education, research and development of technologies that will ensure equality of access to information for all people.

Mr. Paciello has more than 17 years experience in the area of assistive technology and user interface design, including 10 years as Program Manager for Digital Equipment Corporation's Vision Impaired Information Services (VIIS) office and 3 years as Director of the Yuri Rubinsky Insight Foundation (YRIF). At Digital Equipment Corporation, Mr. Paciello produced the computer industry's first mainstream CD-ROM containing computer documentation that was accessible to the blind and visually impaired (VIOLD).

As a professional speaker, Mike has delivered speeches and conducted seminars in the area of web and internet accessibility for clients including Microsoft, Adobe, Sun Microsystems, Easter Seals, Compaq, The Hartford Insurance Company, Yale, MIT, and the National Aeronautics and Space Administration (NASA).

Mr. Paciello received recognition from President William Clinton for creating and launching the Web Accessibility Initiative (WAI) on behalf of the World Wide Web Consortium (W3C) and the White House. He is cofounder of the International Committee for Accessible Document Design (ICADD), and is the creator of WebABLE!, a powerful information Web portal for people with disabilities (http://www.webable.com/).

Part I

The Web and
People with Disabilities

Chapter 1

Introducing Web Accessibility

In This Chapter

- Why make your site accessible?
- Getting to know the disabilities community
- 500 million and growing — a new market niche
- Applying the golden rule
- Reaching the third wave

Chapter 1 introduces you to the world of Web accessibility involving people with disabilities. You will see how the level of awareness has been raised regarding the information needs of people with disabilities and how they are directly affected by elements of the World Wide Web that are currently inaccessible to them.

In this chapter, you will get a sense of the world's population of people with disabilities and better understand their user needs in the information society. In turn, this information will help you appreciate how developing an accessible Web for them now will ensure their continued use of it in the future.

Why Make Your Site Accessible?

The World Wide Web (WWW, or simply the Web) has long surpassed original predictions that it would be "the next killer app" of the Internet. What started out as the home of computer gurus is now an integral part of human society. The Web has become a commodity that everyone has to have and everyone needs to use because it is built upon the most important commodity of the next millennium: information.

Beginning with the launch of the *Web Accessibility Initiative* (or *WAI*, pronounced "way") in April 1997, it became clear that building and redesigning the Web to be accessible to people with disabilities would become an important directive of the *World Wide Web Consortium (W3C)*. Tim Berners-Lee, director of the W3C and inventor of the Web, launched the WAI with the following statement:

> Worldwide, there are more than 750 million people with disabilities. As we move towards a highly connected world, it is critical that the Web be usable by anyone, regardless of individual capabilities and disabilities. The W3C is committed to removing accessibility barriers for all people with disabilities — including the deaf, blind, physically challenged, and cognitive or visually impaired. We plan to work aggressively with government, industry, and community leaders to establish and attain Web accessibility goals.

The Web Accessibility Initiative was launched during the sixth International World Wide Web Conference in 1997. The WAI's (http://www.w3.org/WAI/) mission is as follows:

> The W3C's commitment to lead the Web to its full potential includes promoting a high degree of usability for people with disabilities. The Web Accessibility Initiative (WAI), in coordination with organizations around the world, is pursuing accessibility of the Web through five primary areas of work: technology, guidelines, tools, education and outreach, and research and development.

The World Wide Web Consortium (http://www.w3.org/) is the international industry consortium whose mission is "to lead the World Wide Web to its full potential by developing common protocols that promote its evolution and ensure its interoperability."

Lending his support to the WAI, U.S. President William Clinton stated: "Given the explosive growth in the use of the World Wide Web for publishing, electronic commerce, lifelong learning and the delivery of government services, it is vital that the Web be accessible to everyone."

If you are a person with a disability, no doubt the Web is just as important to you as your "able-bodied" neighbor or coworker. In fact, one could easily argue that the Web is more important to you because it provides access to services, products, and information that are not as easily obtained by you because of circumstances related to your disability. If you are not able to walk or cannot be easily transported from your house to the local Best Buy electronics superstore, you can go to `http://www.bestbuy.com/` on the Web and purchase a new digital television without having to leave your home. Web technology has advanced to the extent that you can do this and many more things quickly and safely.

On the other hand, you don't need to have a disability to understand the advantages of the Web. An individual without a disability is just as easily motivated to purchase products and services online. The convenience factor alone is enough reason to shop online.

So what's the problem? Simply this: Where accessibility and usability of the Web are concerned, there are distinct advantages for able-bodied people over people with disabilities. Common Web tasks such as reading, searching, and purchasing are often difficult, or in some cases impossible, for a person with a disability to perform. Many Web sites are not accessible to large segments of the disability communities — particularly people who are blind, deaf, or hard of hearing. As the interactive nature of the Web continues to expand, those with physical disabilities or speech disabilities may have trouble with immersive virtual reality systems that require walking, reaching, and grasping, or human-to-computer voice response systems that require clear speech.

Not convinced? Think I'm exaggerating? Turn off image loading in your Web browser and spend one hour surfing the Web. Visit your favorite sites and bookmarks. Peruse anywhere you want. And don't be satisfied with just viewing the sites' home pages. Surf as you normally would, going down into the site at least one or two levels. I guarantee that you'll find it extremely challenging because most of the sites you visit have not consistently implemented the simplest of all accessibility attributes: the `alt` (alternative) text attribute to the HTML element `IMG` (image). The result is a Web page that is extremely difficult to navigate, particularly for people who are blind. If the `alt` text were present, it would replace the image, providing the user with the same information a sighted person receives.

So whose fault is it? The Web engineering community because it failed to recognize the need early on in the Web development cycle? Web site designers and content creators because they have not taken time to familiarize themselves with appropriate accessibility coding and design? Standards organizations because they have not enforced or implemented standards that ensure the accessibility of the Internet or the Web? Industry because their focus is on revenue and not product usability? Disability organizations and assistive technology corporations because they are not able to keep pace with emerging technology? There's no rocket science here — everyone shares a measure of the responsibility.

My point is not to focus on the problems, but to identify the solutions. Many do exist. Where solutions do not exist, the goal is to build enough awareness about the issue so a solution can be developed. The primary objectives of this book are to thoroughly educate you (the designer, the developer, and the user) about accessibility issues and to present the solutions required to make the Web accessible — that is, WebABLE!

Since the WAI launch, hundreds of individuals, organizations, and businesses all over the world have swarmed to support the mission of the WAI and its program office. Perhaps the most amazing aspect of their support is that most of it is donated time — volunteer work that is totally pro bono.

That's right, free labor. Why? What motivates so many people to want to take on a seemingly impossible task? Actually, there are several very good reasons that start with the community of people with disabilities themselves. Who are they? Why are things so difficult for them? How is the Web inaccessible to them?

Getting to Know the Disability Communities

The subject matter in this book centers on people with disabilities. It is much easier to explain what you need to do to make your Web site more accessible when administrators, designers, and engineers understand the user characteristics of the disabled. If you are not a person with a disability — for example, loss of vision, hearing, or mobility — then likely you're not familiar with their needs as Web surfers. In the development of any interface, the first rule of thumb is "know thy user."

Note

The following descriptions of the different groups of people with disabilities are not intended to be complete or exhaustive. For the purposes of this book, I try to identify the communities of people with disabilities who appear to be most affected by the inaccessibility of the Web today. Therefore, in addition to descriptions of the disability, each category also briefly highlights the barriers that people with disabilities must overcome in order to use the Web.

This does not imply that other communities are, in some way, affected less or not affected at all. Remember that the goal is to achieve accessibility to the greatest extent possible for all people with disabilities. By building awareness around the current issues and the most visible areas of inaccessibility, this book helps identify new areas that can be resolved in the near future.

People Who Are Blind or Visually Disabled

Of all the disability communities concerned by the inaccessibility of the Web, people with visual disabilities probably rank first. This is primarily due to the graphical nature of the Web's client-server interface.

Visual disabilities vary in category including low vision, color blindness, and total blindness. The following sections contain a description of each.

Low Vision

The American Academy of Ophthalmology (http://www.eyenet.org/aao_index.html) defines low vision as follows:

> If ordinary eyeglasses, contact lenses or intraocular lens implants don't give you clear vision, you are said to have low vision. Don't confuse this condition with blindness. People with low vision still have useful vision that can often be improved with visual devices. Whether your visual impairment is mild or severe, low vision generally means that your vision does not meet your needs. Using visual devices to improve your vision usually begins after your ophthalmologist has completed medical or surgical treatment or determined that such treatments will not improve your vision.

On the Web and when using computers, many people with low vision use specialized monitors or software that increases the size of text or images large enough for the individual to see. Web sites that use absolute font sizes make it difficult for the low vision user to make this adjustment using his or her computer.

Additionally, some low vision users have difficulty making out certain font styles. Italic text, for example, may be difficult for a low vision user to read without assistive software. (However, in all honesty, italic text is difficult for individuals with good vision to read. This is often due to inadequate screen resolution or poor font quality.)

Color Blindness

People who are color blind often have difficulties in distinguishing between combinations and/or pairs of colors.

On his Web site, (`http://www.delamare.unr.edu/cb/`), Andrew Oakley provides in-depth descriptions of the various types of color blindness:

> At the back of your eyes you have Cones and Rods. Cones pick up colour. Rods pick up brightness. There are blue cones, red cones and green cones. They pick up different wavelengths of light. Colour blind people have less numbers of particular cones than normal, so they get colours confused. Some people are more colour blind than others.

Web accessibility issues for individuals who are color blind often involve color combinations that are not properly coordinated or do not provide high contrast. Images without alternative text are also an inconvenience, particularly when the individual is not able to discern what the image is due to the nature of his or her blindness.

You can learn more about color blindness at the Lighthouse International Web site at `http://www.lighthouse.org/color_contrast.htm`.

Blindness

Blindness comes in a variety of degrees. Most people defined as being blind often do have a measure of sight, as limited as it might be. For example, a person whose level of sight is equal to or less than 20/200 — even with corrective glasses or lenses — is considered *legally blind*. A person who is completely sightless is considered to be *blind*.

Many diseases and conditions contribute to or cause blindness including cataracts, cerebral palsy, diabetes, glaucoma, and multiple sclerosis. Many of these conditions are more prevalent as we age.

Web accessibility for people who are blind is a considerable challenge based on the obvious fact that the Web is a visual interface. Images without associated text, frames, tables, forms, and interactive content are just a few of the problems that perplex these users.

People Who Are Deaf or Hard of Hearing

Up front, it's very important that so-called "able-bodied" people understand some important distinctions regarding those who are typically classified as "people with disabilities." For example, this section purposely differentiates between people who are *deaf* and people who are *hard of hearing*. This is a crucial distinction. Generally speaking, the deaf do not consider themselves hard of hearing. Their hearing is not impaired; it simply does not exist.

Quite obviously, then, a person who is hard of hearing is one who has lost a degree of his or her ability to hear. These individuals may need an amplifying device in order to have functional hearing.

It is also important to know that people who are deaf do not consider themselves "disabled" or "functionally limited." They prefer the distinction of being their own culture that includes their own form of communication, which is usually sign language. Deaf World Web (http:// www.deafworldweb.org/asl/) is an excellent resource that provides information about the deaf culture and includes a useful American Sign Language dictionary.

The fact that the deaf culture is included in the category of people with disabilities is primarily based on the increasing prevalence of Web multimedia content that includes dialogue and sound but does not include captioning. Additionally, with the growing popularity of speech recognition interfaces, people within the deaf culture who have limited speech capacity (or none at all) run the risk of being shut out of next-generation computing interfaces all together.

People with Speech Disabilities

Individuals who have speech limitations or speech disabilities collectively include a population of people who have weakened speaking ability or a complete loss of their ability to speak.

You may not consider the population of people with limited speaking ability to be that significant. In fact, there are a variety of disabilities and conditions that include limited speech functionality as a secondary aspect of the disability. The publication titled *Extend Their Reach* notes the following:

> Speech limitations, like other disabilities vary greatly in severity and cause. They might result from severe language delay, cerebral palsy, mental retardation, autism, traumatic brain injury, or stroke. Speech problems can also result from several disorders affecting nerves and muscles including ALS, dystonia, Huntington's disease, multiple sclerosis, and muscular dystrophy.

Similar to the problems facing people who are deaf and hard of hearing, individuals with speech disabilities are dangerously at risk of being ignored as speech recognition interfaces become the norm.

People with Physical Disabilities and Motor Impairments

For people with physical disabilities or motor impairments, accessibility issues can take on a wide range of challenges. Some people have use of their hands while others do not. Some have the ability to use mouth sticks and head pointers while others rely on infrared devices.

Physical impairments are wide and varied. They include conditions such as muscle weakness, paralysis, joint discomfort, and spinal injuries, or disease processes such as arthritis and muscular dystrophy.

Functional limitations as a result of Repetitive Strain Injury (RSI) have increased dramatically over the years. Ironically, one of the key reasons for this increase is directly related to use of personal computers. This led to a growth in the use of emerging technologies such as speech recognition.

The growing popularity of Web appliances and devices such as WebTV and Web kiosks, if not properly designed and tested, will present numerous challenges for the physically challenged user.

People with Cognitive or Neurological Disabilities

Cognitive and neurological disabilities may seem a little more difficult to address. However, as with outwardly apparent physical disabilities, the improvements made to your Web authoring techniques will serve more than the disabilities community.

Individuals with dyslexia, dyscalculia, and auditory perception difficulties benefit from information being presented in short, discrete units. Easily digestible chunks of data make the important points in your content stand out as well.

Some neurological conditions can result in users being sensitive to excessive flashing in animations or blinking that occurs within certain ranges of frequencies. Seizure disorders have been known to be triggered by such events. Any time that the eye is distracted from the real content of the page, your meaning may be lost.

500 Million and Growing — A New Market Niche

At first glance, it's hard to believe that more than 500 million people make up what has always been viewed by mainstream industry as a small niche market. This is in fact a 1980 estimate maintained by the United Nations and contained in their report on the World Programme of Action Concerning People with Disabilities. No doubt this number has increased in 20 years. Of course, not all of those people are impacted by accessibility issues on the Web. It's important to remember, however, that people with disabilities are found in all socio-economic levels. Therefore it's likely that the number of disabled users with access to the Internet will be proportionally similar.

Even among Internet users who have disabilities, not all of them present accessibility issues. For example, someone who is paraplegic will likely not have trouble typing, operating a mouse, seeing, or hearing, unless they have an additional unrelated disability. Someone who must refrain from strenuous aerobic activity due to a heart condition won't necessarily have trouble surfing the Web.

This doesn't mean, however, that the community of people with disabilities that do impact Internet use isn't statistically meaningful. Indeed, this community numbers in the tens of millions.

As more and more people gain access to the Internet, the wired disability community continues to grow at incredible rates.

Note the following statistics:

- In the European Community, approximately 37 million people (or 1 in every 10 citizens) have a disability.

- There are more than 4 million Canadians (or 1 in every 7 individuals) with a disability.

- Approximately 3.7 million Australians live with a disability.

- In the United States, according to statistics available in the 1997 U.S. census, approximately 54 million people live with at least one disability.

There's no hiding the fact that the population of able-bodied people outnumbers the population of those with disabilities. This remains true on the

Internet. On the other hand, tens of millions is an enormous number, regardless of what the population is for the rest of the world. Can you think of any business that wouldn't jump at the opportunity to position and sell a product or service to as little as 1 percent of that number? With the growth of electronic commerce making it easier for businesses to reach the consumer, who could afford to miss the opportunity? A report posted in *Internet World* based on data compiled by Jupiter Communications forecasts that online consumer spending will reach $29.4 billion by the year 2002.

Perhaps best of all, the Web enables you to market and sell your products to the disabilities community market with minimal effort and cost. In most cases, it's no more work than simply ensuring that your Web site includes a text description of the products and services you are selling. E-commerce is the new wave for businesses. In one fell swoop, you can go online, reach bigger markets, and establish a new clientele.

The purchasing power of people with disabilities is also at an all-time high. According to a 1998 report released by the President's Committee on Employment of People with Disabilities, "Consumers with disabilities control more than $175 billion in discretionary income. They, like all consumers, are more likely to patronize businesses where they feel welcome. Accessible stores, products and services, along with employees with disabilities, will help customers with disabilities feel that their business is appreciated."

Clearly, the effect of the population of consumers with disabilities on electronic commerce has influenced world governments and international standards–based industry organizations to launch creative initiatives to address the needs of the disabled consumer. Exactly how large is the market?

National Statistics

In December 1997, based on a census taken during the four-month period of October 1994 to January 1995, the U.S. Department of Commerce's Bureau of Census reported that 1 in 5 Americans or (54 million people), have some kind of disability. This is about 20 percent of all U.S. citizens, which comprises a larger minority population than African Americans (approximately 30 million).

Additionally, the same report highlighted the following breakdown of those same statistics:

- 1 in 10 Americans has a severe disability.
- Among American children aged 6–14, 1 in 8 have some type of disability.

- 1 in 2 Americans 65 years and older has a disability.
- 1 in 5 Americans between the ages of 15 and 64 years has a disability

To further emphasize that nearly 20 percent of all Americans have some kind of disability, InfoUse's 1996 edition of the *Chartbook on Disability in the United States* (http://www.infouse.com/disabilitydata/chartbook.choices.html) estimated that 19.4 percent of noninstitutionalized U.S. citizens had a disability at that time.

Of those, an estimated 24.1 million people had a severe disability. The report estimated that 13.1 million people use assistive technology for anatomical, mobility, hearing, vision, and speech disabilities

Another interesting series of data related to people with disabilities involves the workforce. The *Chartbook on Disability in the United States* (1996 edition) noted that in 1994, 52 percent of the disabilities population were part of the workforce. Continuing advances in assistive technology are enabling more people with disabilities to enter the workforce in greater numbers than ever before. The Internet and Web have played important roles in this advancement, simply by providing greater access to information and services to all people.

While the statistics regarding the population of people with disabilities may come as a surprise to some, almost 30 percent of all families in the United States are affected by a member who has some type of disability. The report titled *Families with Disabilities in the United States,* released in 1996, noted the following:

- An estimated 20.3 million families, or 29 percent of all 69.6 million families in the United States, have at least one member with a disability (as measured by having an activity limitation).

- When a family has a member with a disability, that member is most likely to be a householder. For example, in 88 percent of partnered families with disabilities, one or both partners have a disability.

- An estimated 2.3 million partnered families contain one or more children with a disability.

- Some 3.8 million families, or 6 percent of all families, contain one or more children with disabilities. Most of these (3.4 million, or 89 percent) have one child with a disability.

- The rate of disability is 29 percent for white families, 32 percent for black families, and 22 percent for other races. Among Hispanic families, 23 percent have members with disabilities.

Cross-Reference

You can find the *Families with Disabilities in the United States* report at http://dsc.ucsf.edu/UCSF/.

Almost without a doubt, very few families in the United States are left untouched or unattached to an individual with a disability.

International Statistics

On the international front, the statistics are not as easy to come by, but those that are available are consistent with the United Nation's report that approximately 1 in 10 people in the world has a disability.

In Europe, 1 in every 10 people, or approximately 37 million people, have a disability. The European Community (EC) is committed to the mainstreaming of their citizens with disabilities. In a report distributed by the European Commission on the Equality of Opportunity for People with Disabilities, the Commission reported that 5.5 billion ECU would be allocated during the five-year period of 1994–1999 to combat exclusion of people with disabilities in the workforce.

This report also noted that the Employment-HORIZON initiative, a part of the EC's Employment Community Initiative, was allocated 513 million ECU to advance employment opportunities for people with disabilities.

Again, to further emphasize the high level of awareness involving issues of accessibility to information, the Commission on Equality of Opportunity stated the following in Part VI of "Mainstreaming: Information and Communication Technologies (ICTs)" (emphasis is mine):

49. The Commission is actively interested in exploring the possibilities for **harnessing all aspects of the Information Society in the achievement of equal opportunities for people with disabilities** and in improving their living and working conditions. These questions are discussed in general in the Commission's Green Paper on Living and Working in the Information Society: "People First". An internal ad hoc group will be set up by the Commission to take this forward with the mandate to examine the scope for a special initiative at European level, building on relevant experience to date, for example in the TIDE Initiative. This will be based on a review of good applications of Information and Communication Technologies (ICTs) in favour of people with disabilities, and the further potential for developing economies of scale in making ICTs more widely accessible and useful to people with disabilities.

A 1996 report published by the Canadian Premier's Council on the Status of Disabled Persons in Canada indicated that of the 27.3 million Canadians, about 1 in 7 (or approximately 4.2 million) have a disability. In Australia, according to a 1997 publication by the Australian Institute of Health and Welfare, the Australian Bureau of Statistics (ABS) reported in 1993 that there were 3,176,700 people in Australia with a disability. This constituted approximately 18 percent of the Australian population at that time.

Applying the Golden Rule

Now that Web accessibility issues are gaining mainstream attention, some companies are beginning to wake up and adjust their development strategies to include accessible user features. In the past, Web accessibility for people with disabilities was an afterthought. At best, corporations would support an adaptation or a fix to the interface or application after the finished product was released to the public.

The basis for this strategy often is dictated by inadequate market studies, limited user focus, and the perception that a small population of people with disabilities actually use the Web. However, as we have seen from the statistics previously referenced, approximately 1 in 10 Americans has a disability that could impact their Web surfing experience.

There are other compelling reasons for designing Web accessibility today, including government legislation and international technical standards. A very practical approach (and one that I have personally found to be quite influential) toward motivating anyone is to hit them where they'll feel it the most — in the heart. Because statistics indicate that almost 30 percent of all families living in the United States include someone with a disability, there's no doubt that the percentage of those affected is much higher when you include acquaintances and friends.

One of the greatest privileges I've had in my professional life was to meet and briefly work with the internationally known linguist Dr. Tony Vitale. Tony was instrumental in the development of the DECtalk synthetic speech synthesizer for Digital Equipment Corporation (since bought and absorbed by PC giant, Compaq Computer Corporation). DECtalk is still one of the most widely used speech synthesizers by people with disabilities. Blind people use DECtalk to read the contents of their computer screens. Individuals who are not able to speak use DECtalk to speak for them. Children with learning disabilities have DECtalk read stories to them. Professor and physicist Stephen Hawking's voice uses the DECtalk speech technology. You can see why Tony is generally revered in the disabilities community for his significant contribution.

Suddenly and with little warning, Tony was diagnosed with Amyotrophic Lateral Sclerosis (ALS), more commonly known as Lou Gehrig's disease. ALS is a fatal, neuromuscular disease that causes rapid deterioration of motor cells in the brain and spinal cord, ultimately leading to severe impairment of mobility, speech, and respiratory functions (Stephen Hawking also has ALS). Tony's health deteriorated quickly, and within months, he no longer was able to speak without the aid of a speech synthesizer. The very device that he was responsible for developing became essential to his ability to communicate with others.

The irony in this incredible turn of events is heart-wrenching to say the least. Perhaps more eye-opening are the statistics that indicate situations like Tony's are not nearly so uncommon as you may have thought. For example, a 1991 report by the U.S. Congress House Committee on Ways and Means stated that individuals between the ages of 35 and 65 have a 30 percent chance of experiencing a disability.

The report also noted that a person's greatest earning potential occurs between the ages of 35 and 65. Gives new meaning to the term *disability insurance,* doesn't it?

When you see the world through someone else's eyes, it's much easier to appreciate why so many Web engineers have made it their mission to improve the quality of the Web for people with disabilities. The biblical principle "Do unto others as you would have them do unto you" applies with equal measure, especially in high-tech circles. In technical jargon, the golden rule is evident through *usability,* or designing for the user. When you design for the user, particularly for the user with a disability, you are doing the right thing.

Do the Right Thing

Let's review "doing the right thing" from the user's perspective. Take a look at your Web site. Is it filled with images, image maps, tables, frames, multimedia, and programming scripts? If so, have you taken the appropriate steps to ensure that each mechanism is accessible? Does your Web site support style sheets? Have you ever included blind or deaf people in your Web site review process? Do you have a Web site review process? Are you HTML 4.0 compliant? If not, why not?

Tip

You can easily check your Web site for most accessibility errors and conditions by using a number of publicly available tools like Bobby (http://www.cast.org/bobby/), LIFT (http://www.usablenet.com/), WebSAT (http://zing.ncsl.nist.gov/webmet/sat/websat-process.html), and WHAT (http://cmos-eng.rehab.uiuc.edu/what/). LIFT and Bobby are also available through WebABLE (http://www.webable.com/).

Additionally, the W3C provides an HTML 4.0 validation service (http://validator.w3.org/) that verifies proper HTML accessibility coding.

Cross-Reference

Please refer to Chapter 7 for additional information about Web accessibility and HTML validation services.

The point is that, barring copyright or privacy rules, isn't access to content by the largest number of people possible the vision of the Web? *PC Computing* magazine put it succinctly: "Your data will be everywhere." The whole driving force of the Web is to get your content and Web services to the public, on a global basis. Why wouldn't you design and create that information for the widest population, including people with disabilities? Seems to me that 500 million is a reasonably-sized market, so there are at least 500 million good reasons for doing it.

The Third Age

Remember too that the Third Age generation (the elderly) comprises the largest and fastest growing population segment in the world. Contrast the demographics of a report generated by the World Health Organization (WHO) in 1997 (Figure 1.1) with the projected increase of the elderly population by the year 2025 (Figure 1.2). In several nations, nearly 20 percent of the population will consist of elderly people!

Figure 1.1   WHO demographics for population of people 65 and older in 1997

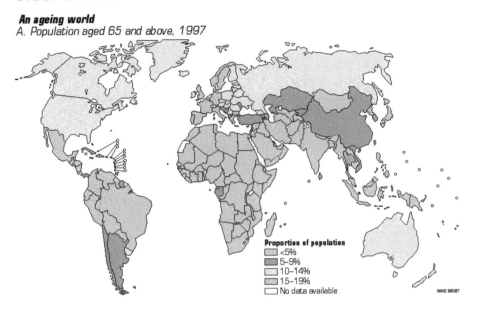

An ageing world
A. Population aged 65 and above, 1997

Proportion of population
<5%
5–9%
10–14%
15–19%
No data available

WHO 98087

Figure 1.2   WHO demographics for population of people 65 and older by 2025

B. Population aged 65 and above, 2025

Proportion of population
<5%
5–9%
10–14%
15–19%
20%+
No data available

WHO 98088

The Third Age generation is also made up of a significant population of individuals with *multiple disabilities* — people affected by more than one type of disability. As people age, it is quite common to experience loss of sensory combinations including vision, hearing, and mobility.

Regarding the elderly and causes of disability, WHO also reported the following:

- Cancer and heart disease are more related to the 70–75 age group than any other; people over 75 are more prone to impairments of hearing, vision, mobility, and mental function.

- Over 80 percent of circulatory disease deaths occur in people over 65 years of ago. Worldwide, circulatory disease is the leading cause of death and disability in people over 65 years of age.

In spite of these statistics, make no mistake about it: The Third Age generation is quite adept at using the Web. Web sites set up and maintained for and by senior Web users abound. Just take a peek at Web sites for organizations like the American Association of Retired Persons (`http://www.aarp.org/`) and the Third Age home page (`http://www.thirdage.com/`).

Lack of Awareness

My personal experience has been that most Web site designers will make their sites accessible when (and if) they know how. I spend more time delivering seminars, workshops, and conference presentations about web accessibility than other single task I can think of.

However, understanding the problem is only half the battle. Misconceptions abound. I believe that at least 50 percent of the problem involving Web accessibility is lack of awareness. Most people involved with the Web, whether as developers or users, simply are not well informed when it comes to accessibility.

Preparing for the Future

What about the future Web? The current technology buzz involves computers and Web browsers that feature speech input/output subsystems. Now you can talk to your computer using voice commands rather than typing them. Today, new cars are being built with Web browsers and navigational systems that feature speech-based interfaces.

New technology? The next wave? Hardly! People with disabilities have relied on this technology for years. Speech technology is a primary tool in the adaptive and assistive technology communities. These "next-generation products" were designed to aid the blind, physically challenged, and non-verbal. Suddenly what was an assistive technology is now a mainstream consumer electronic device.

Everyone, Everywhere

Let's not forget that the challenge of Web accessibility is of great importance on the international front — particularly where awareness is concerned. In the past two years, primarily due to the work coming out of the Web Accessibility Initiative (WAI), several organizations in Europe, Asia, and Australia have examined the complex issues involving internationalization and Web accessibility for the disabilities population. WAI's work resulted in the implementation of country- and community-specific guidelines as well as new products and tools. No doubt, this fervor will spread to the remaining continents in the near future.

One example of the effect Web accessibility is having on the international front is highlighted by the collaborative effort accomplished by consortia members of the HARMONY (Horizontal Action for the Harmonisation of Accessible Structured Documents) Project (`http://www.esat.kuleu-ven.ac.be/teo/docarch/projecten/harmony/harmony.en.htm`). HARMONY is a funded program developed by the Technology Initiative for Disabled Elderly People (TIDE). TIDE is also a valuable funding resource for the WAI in Europe.

HARMONY's chief goal as stated in their final report is as follows:

> To increase the *quantity and quality* of information accessible to "print-disabled" people — especially in daily newspapers — by stimulating the publishing community via a process of involvement, lobbying and standardization, and by encouraging them to adapt their existing electronic production systems to make use of appropriate new document-structuring concepts.

Initially, the HARMONY Project started out as an initiative concentrating on print-based material. However, because of the overwhelming prominence of the Web, the project strategy was extended to include the Web. Implementation of that strategy resulted in significant accessibility enhancements for newspapers, publishers, information providers, and specialized print services throughout the European Community.

Reaching the Third Wave

The fact that the Web is inherently inaccessible is not the result of some malicious or premeditated intent. The Web followed a very typical development process based on standard engineering processes that, all too often, do not include considerations for people with disabilities. Web page designers and content producers observe similar methods. Subsequently, most advanced technologies are not accessible to people with disabilities. Until now, it was satisfactory to create an assistive or adaptive device (or application). Until now, very few laws or standards mandated accessibility.

The Web is a growing, pervasive phenomenon. It is a global technology that requires programmers to think in holistic terms, fully understanding the number of people who are directly affected by current Web barriers and redesigning Web interfaces to adapt to the user. This is the essence of true *personalization* — Web design that ensures accessibility for every user by adapting to the user's preferences.

To clarify, I'm not talking about *customization,* which is the ability of a Web surfer to visit a particular Web site and custom start his or her own page. In essence, that's no more powerful than being able to adjust the desktop on your PC. Rather, a personalized Web, based on recognized user characteristics or preferences, serves content of any construct to users in any form they like.

The personalized Web service is dynamic and pervasive. Regardless of where you are or what you want, you get what you want the way you want it. Even when extracting that data from multiple sources, likely including multiple content types, you get personalized service.

Web personalization is still in the early stages of development and likely, it will be a few years before the average user sees it. In technology, the best way to implement accessibility is through usability. Focus on the user's ability to interact effectively and efficiently with your interface and you'll achieve a greater level of user satisfaction. Include people with disabilities in your user studies and testing. Keep in mind that people with disabilities are not looking for your sympathy. Rather, they encourage Web designers to develop creative solutions that include all users.

Summary

In this chapter, you learned that the number of disabled Web users is far larger than most people might have thought. These individuals have a wide variety of challenges to overcome when accessing Web sites, from dealing with images that may not be seen and audio that cannot be heard to working with hardware that is difficult to manipulate due to physical disabilities. Cognitive and neurological disabilities can limit the user's ability to digest large volumes of information that is poorly organized, or that competes with distracting animation or other visuals.

The next chapter examines laws and policies that encourage and promote measures of information accessibility for people with disabilities. In many ways, these legal standards lead to the development of new initiatives, such as the Web Accessibility Initiative, which focus on access and innovative solutions for people with disabilities.

References

The HARMONY Project. "The Art of Providing Access to Electronic Documents." HARMONY Consortium, 1997.

The Chartbook on Disability in the United States. InfoUse, 1996.

European Commission on Quality of Opportunity for People with Disabilities. *Communication of the Commission on Equality of Opportunity for People with Disabilities — A New European Disability Strategy*. December 1996. Available from http://europa.eu.int/comm/dg05/soc-prot/disable/index_en.htm.

Department of Commerce, Bureau of the Census. *Disabilities Affect One-Fifth of All Americans*. Washington, D.C., December 1997. Available from http://www.census.gov/hhes/www/disable.html.

U.S. Department of Education, National Institute on Disability and Rehabilitation Research. *Families with Disabilities in the United States*. LaPlante, M.P., Carlson, D., Kaye, H.S., and Bradsher, J.E. Disability Statistics Report (8). Washington, D.C., 1996.

President's Committee on Employment of People with Disabilities. "Opening Doors to Ability." July 1998.

United Nations. *United Nations World Program of Action Concerning People with Disabilities*. 1998. Available from http://www.un.org/esa/socdev/diswpa01.htm.

Department of Commerce, Bureau of the Census. "We the Americans: Blacks." Washington, D.C., September 1993.

2

Chapter 2

Legal Requirements, Policies and Standards

In This Chapter

- The Fifth Principle
- Understanding the "Casey Martin Syndrome"
- Using the ADA to enforce Web accessibility
- Section 508 of the Rehabilitation Act
- Additional U.S. policies and standards that promote accessibility
- International policies and standards

This chapter focuses on laws in several countries that have been created in part to aid in the promotion of Web accessibility, specifically on behalf of people with disabilities. You will learn how and why the Americans with Disabilities Act (ADA) paved the way for many of these legislative actions.

You will understand how legal policies and standards in countries everywhere are providing plenty of incentive for creating an accessible Web.

Before discussing legal policies and standards, I'll first review publicly stated goals and principles involving the Internet and Web of the future.

The Fifth Principle

Often, laws and policies are based on the establishment of clearly structured principles and guidelines. In his speech before the International Telecommunications Union during the 1994 World Telecommunications Development Conference in Buenos Aires, Vice President Al Gore discussed the future of the Global Information Infrastructure (GII) and emphasized the need for global community and universal accessibility, stating:

> The development of the GII must be a cooperative effort among government and peoples . . . Let us build a global community in which the people of neighboring countries view each other . . . as partners, as members of the same family in the vast increasingly interconnected human family.

Exactly one year later, the National Information Infrastructure Advisory Council (NIIAC) supported this measure. In their first report, "Common Ground: Fundamental Principles for the National Information Infrastructure," the council established a core set of design principles involving universal access and service. Interestingly, this council included notable high-tech and Internet experts such as Esther Dyson, Mitch Kapor, and John Sculley. Industry representatives from MCI, Microsoft, Disney, Bellcore, AT&T, Silicon Graphics, and CBS were also present. This is at least an indication of industry's interest in the area of accessibility.

Four of the seven principles set clear national goals for guaranteeing Internet and Web access to all people, including access for people with disabilities. These four principles include, in part, the following:

1. A national goal to enable every individual to have access to the NII should be established by the year 2005. This includes defining levels of access and service capabilities. (Principle #1)

2. All individuals should be able to participate as both consumers and producers of information and service on the NII. (Principle #4)

3. Individuals with disabilities should have access to the NII and, therefore, design issues should be addressed as the NII is developed to ensure access for all individuals with disabilities. (Principle #5)

4. Information from all levels of government should be accessible over the NII. (Principle #6)

This national strategy provides powerful motivation for building an accessible Web infrastructure by all Web participants.

The Fifth Principle is well into its development cycle. However, the groundwork for ensuring information accessibility was laid well before the establishment of the NIIAC principles. The first major piece of U.S.-based legislation to enforce information accessibility was the Americans with Disabilities Act (ADA).

Let's examine how the ADA has been used to effectively ensure Internet and Web accessibility for people with disabilities.

Understanding the "Casey Martin Syndrome"

The Internet — and particularly its most visible entity, the World Wide Web — is increasing the public's awareness where accessibility is concerned.

Because of the enormous focus on electronic information creation and dissemination, government and industry standards groups have invested heavily in accessibility measures in order to avoid the potential "Casey Martin Syndrome" — a term I coined that may be applied to lawsuits charging ADA discrimination based on the inaccessibility of critical information. And while the ADA is U.S.-based legislation, the spirit of the ADA exists in legal verbiage across the continents. The Casey Martin Syndrome is very real.

Casey Martin achieved something many people wouldn't have had the guts to do. He challenged and sued an organized public establishment based on his legal rights as an individual with a disability. He challenged that establishment, the Professional Golfers' Association (PGA), because his ability to make a living was severely impeded by the PGA's refusal to accommodate him as a person with a disability. He challenged them and he won big.

The point of interest in all of this is not what Martin did, but rather how he did it (or at least how his lawyers worked it out for him). Martin used the media and key government personalities to support him. During one well-publicized press conference, Casey Martin appeared with Senator Ted Kennedy, who is the father of a son with a disability (Ted Jr., amputee), and former Senator Robert Dole, a previous presidential candidate with a disability. Two of the most influential figures in U.S. politics were standing on either side of Casey Martin — an all-American guy suffering from an extremely painful circulatory disability. Just watching him walk was a painful experience. All he wanted to do was play professional golf and the PGA wouldn't let him because he required a golf cart that they determined would give him an unfair advantage over other golfers. What was about to unfold would end up as a public relations nightmare for the PGA.

The media ate it up. Frankly speaking, the PGA never had a chance. The judge ruled in favor of Martin, hands down. Everyone came to Martin's side, including key industry tour sponsors such as Nike and The Hartford. The PGA suffered a black eye it's still trying to recover from. But it could have been much worse. Martin could have sued the PGA for millions of dollars and walked away with a victory. Public sympathy was overwhelmingly in his favor. Perhaps more than anything else, the publicity of Casey Martin's victory sounded a warning for employers and business owners everywhere. Rules that on the surface seem fair may need to be reconsidered if they require specific physical actions that may be difficult for people with disabilities. The ADA covers more than that, as we'll see, but physical access is one of the easiest concepts for those unfamiliar with the disability communities to understand.

Reviewing the Status Quo

Now, let's apply the same situation to industry. The story goes something like this: One of your employees with a disability is not able to access work-related data stored on your internal Web site — information that is crucial to his or her job performance. The employee sues you based on an ADA violation.

Why? Because you have not provided a reasonable accommodation for the employee.

Your Web pages are not accessible to people with disabilities. The scenario is quite similar to Casey Martin's. The employee's ability to accomplish their work in a productive manner and subsequently earn a living is being compromised by the inaccessibility of the Web site. According to the ADA, every corporation that employs 15 people or more must reasonably accommodate employees. Scenarios like this are just starting to crop up on a small scale.

Soon enough something on a much larger scale will occur. It will only take the initiative, persistence, and highly publicized actions of one person with a disability or an organization representing people with disabilities, to expose the ignorance of a large corporate or government entity. For example, in November of 1999, America Online (AOL), one of the largest internet service providers in the world, was sued by the National Federation of the Blind, the National Federation of the Blind of Massachusetts and nine individuals, charging that "America Online's Internet Service (AOL service) is inaccessible to the blind, thereby violating the Americans with Disabilities Act (ADA)."

In another example, Randy Tamez, who took on the Metropolitan Transportation Commission (MTC) of the San Francisco Bay Area because the MTC created an inaccessible Web site that included important transportation information that Tamez could not access as a person with a visual disability.

Randy subsequently filed a complaint alleging that he was not able to use the MTC's Web pages that included transportation information intended for the public. Because the information was not accessible to a citizen with a disability, the MTC, according to the ADA, was in violation.

In an interview with the *San Francisco Examiner,* Randy noted that, "A Web site is like a public building. You open it up to the public, and you can't discriminate against people who can't get up the stairs."

What happened next? Within days of filing the complaint, publicity of the event was all over the Web — online newspapers, Web sites, electronic mail listservs, and chat groups. Everywhere. Publicity spread like wildfire. In no time, people were writing letters in support of Randy. The very technology that Randy Tamez was suing to use was being used to support him!

Successful cases will often generate similar actions elsewhere in the country. Weeks later, Tamez and a Chicago-area man, Kelly Pierce, both filed complaints with the Federal Transit Administration regarding access problems with the Web sites maintained by their local transit agencies. It will only be a matter of time before things escalate to the point where businesses and offices of public accommodation will realize that the World Wide Web is an important citizen access point and it must be accessible to all people.

The Tamez and Pierce cases are not the only complaints involving Web accessibility on record. As early as 1995, an ADA complaint was filed against the city of San Jose, California citing the inaccessibility of certain city documents. A blind city commissioner who had been appointed by the Mayor and San Jose City Council to advise on disability matters was not able to access and read city council documents posted on the city Web site. The documents were posted in a file format known as Portable Document Format (PDF), a file format which is completely graphical in nature and difficult to render accessibly (refer to Chapter 11 for information about PDF accessibility). The city subsequently made the appropriate adjustments to ensure that versions of files and information stored on their Web site were accessible.

There is an interesting touch of irony in this case that involved Cynthia Waddell, San Jose's ADA Coordinator, herself a person with a hearing disability. In an e-mail exchange between Cynthia and myself, she explained:

> As I researched this issue, I immediately realized that I too was a stakeholder in the issue because I needed captioning of audio and video files posted on the web. Shortly after our adoption of accessibility standards in 1996, the USDOJ policy ruling came out discussing the need for "effective communication" in the design of web pages. As you know, our webmaster standard requires that if a pdf file is posted, an accessible version must also be posted; captioning is also required for audio and video-clip files.

The City of San Jose's Web site (`http://www.ci.san-jose.ca.us/oaacc/disacces.html`) stands as a "model citizen" for the rest of the world to imitate. Much of this can be attributed to the tireless efforts of Cynthia Waddell.

More Legal Victories

The legal agencies in the United States are gathering steam to ensure the enhancement of Web accessibility and information technology rights of people with disabilities. Cynthia's presentation during the American Bar Association conference cited several recent settlements involving Internet and Web accessibility.

For example, in 1996, the Office of Civil Rights (OCR), United States Department of Education discussed the ADA's definition of effective communication based on a complaint filed by a student involving a university that failed to provide accessible Internet access. The OCR settlement letter stated:

> The issue is not whether the student with the disability is merely provided access, but the issue is rather the extent to which the communication is actually as effective as that provided to others. Title II [of the Americans with Disabilities Act of 1990] also strongly affirms the important role that computer technology is expected to play as an auxiliary aid by which communication is made effective for persons with disabilities.

In a later settlement (1997), the OCR defined "effective communication" as consisting of three essential components: "timeliness of delivery, accuracy of the translation, and provision in a manner and medium appropriate to the significance of the message and the abilities of the individual with the disability."

It is worth noting the potential pervasive effects these settlements will have on the Web, Internet, National Information Infrastructure and, very likely, the Global Information Infrastructure. Each of these entities is viewed as a critical communication interface to the next generation of business, education, and medical industries. In the United States, if you are in the information technology business, you will be held accountable for delivering accessible data.

In another ruling (*Tyle vs. City of Manhattan*), OCR emphasizes that the U.S. courts have held that ADA-defined public entities are violating ADA obligations if the entity only responds to accommodation requests on a person-by-person basis. ADA obligates public entities to develop policy in advance of any accommodation request for accessibility aids or services.

Services? Absolutely. And, with the growing prevalence of Web-based customer and government services, this is sure to lead to the next layer of legal settlements and rulings that guarantee access to people with disabilities. It is my opinion that recently enacted U.S. policy and legislation, including Section 508 and the Telecommunications Act, will further increase this level of responsibility for most public entities. (I discuss both Section 508 and the Telecommunications Act in later sections.)

To further support my argument, note again how the Office of Civil Rights continued in their 1997 settlement letter. Highlighting the responsibility of all public entities to provide accessible software and hardware, OCR stated, "the subsequent substantial expense of providing access is not generally regarded as an undue burden when such cost could have been significantly reduced by considering the issue of accessibility at the time of the initial selection." If you are a public entity as defined by the Americans with Disabilities Act, you are required to ensure that barriers are removed and new barriers are not constructed.

Perhaps the most difficult aspect of this right-to-information jargon is the public perception that people with disabilities are asking the impossible. People with disabilities aren't asking the impossible — they are simply requesting access to information and services that will help them function more productively in society. The focus is on enhancing what is currently available on the Web.

Note

For additional information about legal cases involving accessibility on the Web and Internet, I recommend you read Cynthia Waddell's paper, "The Growing Digital Divide in Access for People with Disabilities: Overcoming Barriers to Participation." You can find an online copy of the paper at `http://www.digitaleconomy.gov/`.

Enhancing, not protecting. Protection is not what's at stake. What's at stake is empowerment and independence. The right to support oneself and one's family. The ability to experience working with next-generation technology that is already accessible. These are the kind of things individuals believe are worth fighting for. Lawyers are catching the wave and using situations like those of Casey Martin and Randy Tamez to drive home the seriousness of Web accessibility. Web sites like the Internet Lawyer (www.internetlawyer.com/index.htm) and ADA Technical Assistance (www.adata.org) already are offering immediate online legal advice.

Using the ADA to Enforce Web Accessibility

The Americans with Disabilities Act (ADA) was signed into law in 1990 during the administration of former President George Bush and officially went into effect in 1992. The ADA includes several provisions on behalf of people with disabilities. Included as part of its mandate are guarantees of "reasonable accommodation" and "effective communication," particularly in areas of employment, public services, public accommodations provided by private entities, and telecommunication services. Until recently, reasonable accommodation generally was the topic that employers and public accommodation providers were concerned with. The popularity of the Internet and the World Wide Web has dramatically changed that focus.

To establish the need for the ADA, the United States Congress first recorded established facts involving the disabled. These findings are posted as an introduction to the Act. Notable among them are the following (emphasis is mine):

- Some 43,000,000 Americans have one or more physical or mental disabilities, and **this number is increasing** as the population as a whole is growing older.

- Discrimination against individuals with disabilities persists in such **critical areas** as employment, housing, public accommodations, education, transportation, **communication**, recreation, institutionalization, health services, voting, and access to public services.

- Unlike individuals who have experienced discrimination on the basis of race, color, sex, national origin, religion, or age, individuals who have experienced discrimination on the basis of disability have often had **no legal recourse** to redress such discrimination.

- Individuals with disabilities continually encounter various **forms of discrimination**, including outright intentional exclusion, the discriminatory effects of architectural, transportation, and **communication barriers,** overprotective rules and policies, failure to make modifications to existing facilities and practices, exclusionary qualification standards and criteria, segregation, and relegation to lesser services, programs, activities, benefits, jobs, or other opportunities.

- The Nation's proper goals regarding individuals with disabilities are to assure equality of opportunity, full participation, independent living, and economic self- sufficiency for such individuals.

- The continuing existence of unfair and unnecessary discrimination and prejudice denies people with disabilities the opportunity to compete on an equal basis and to pursue those opportunities for which our free society is justifiably famous, and costs the United States billions of dollars in unnecessary expenses resulting from dependency and nonproductivity.

As you can see, the findings specifically highlight *communication barriers* as a critical area and form of discrimination affecting people with disabilities. Four years after the ADA became mandatory a question surfaced involving the interpretation of this area.

Finding More Information About the ADA

You can read the full text of the ADA on the Web at `http://jan-web.icdi.wvu.edu/kinder/pages/ada_statute.htm`.

Several informative Web sites provide valuable services involving the ADA. The Americans with Disabilities Act Document Center (`http://janweb.icdi.wvu.edu/kinder/`), maintained by Duncan Kinder, is one of the most informative ADA-oriented Web sites available. This site provides a complete set of ADA-related documentation including reference sheets and explanations of each title.

The ADA Technical Assistance Program (`http://www.adata.org/`) is a federally funded program that provides information, training, and technical assistance to businesses and people with disabilities about ADA compliance.

The Alexander Law Firm provides an online brochure called "The Americans with Disabilities Act Questions and Answers" that you can find at `http://consumerlawpage.com/brochure/disab.shtml`.

The Communications Workers of America publishes an ADA Frequently Asked Questions (FAQ) Web page at `http://www.igc.apc.org/cwatx/ada.html`.

The Web Gets Involved

In 1997, *The National Disability Law Reporter* published an inquiry that U.S. Senator Tom Harkin made in 1996 on behalf of a constituent.

The substance of that request involved understanding what extent the ADA required Web pages to be accessible to people with disabilities. The Justice Department responded as follows:

> The Americans with Disabilities Act (ADA) requires State and local governments and places of public accommodation to furnish appropriate auxiliary aids and services where necessary to ensure effective communication with individuals with disabilities.
>
> Covered entities under the ADA are required to provide effective communication, regardless of whether they generally communicate through print media, audio media or computerized media such as the Internet. Covered entities that use the Internet for communications regarding their programs, goods, or services must be prepared to offer those communications through accessible means as well.
>
> ...examples include providing the Web page information in text format, rather than exclusively in graphic format. Such text is accessible to screen reading devices used by people with visual impairments. Instead of providing full accessibility through the Internet directly, covered entities may also offer other alternate accessible formats, such as Braille, large print, and/or audio materials, to communicate the information contained in web pages to people with visual impairments. The availability of such materials should be noted in a text (i.e., screen-readable) format on the web page, along with instructions for obtaining the materials, so that people with disabilities using the Internet will know how to obtain the accessible formats.
>
> The Internet is an excellent source of information and, of course, people with disabilities should have access to it as effective as people without disabilities.

The response left no doubt in the minds of anyone — the ADA requires Web accessibility.

As she has so many times emphasized, Cynthia Waddell, ADA Coordinator for the City of San Jose, California, and an expert in the field of ADA compliance and Web accessibility, in a presentation before the American Bar Association concluded the following:

> Without the application of ADA requirements to the Internet, new barriers to effective communication and global commerce will be erected that will have a discriminatory impact upon individuals with disabilities. Accessible Web design should be mandated so that everyone, regardless of age or disability, or the limitations of their computer equipment, can participate in the benefits of the World Wide Web.

The Americans with Disabilities Act is one example of U.S. legislation that requires Web accessibility in certain settings, but it clearly does not stand alone. Without a doubt, Section 508 of the Rehabilitation Act may be the single most important piece of U.S. legislation that unequivocally addresses the needs of people with disabilities in the field of advanced information technology.

Section 508 of the Rehabilitation Act

Section 508 of the Rehabilitation Act is a vital piece of legislation that defines the processes used by the federal government to procure electronic and information technology. Section 508 (as it is most commonly referred to) was amended in August 1998 to more clearly regulate the federal procurement process and ensure compliance by assigning the monitoring role to the Department of Justice.

An important focus of the Act is to ensure access to electronic and information technology made available to people with disabilities that are federal employees or members of the general public.

Section 508 establishes the following requirements for federal agencies and departments:

(a) Requirements for Federal Departments and Agencies:

(1) Accessibility:

(A) Development, procurement, maintenance, or use of electronic and information technology: When developing, procuring, maintaining, or using electronic and information technology, each Federal department or agency, including the United States Postal Service, shall ensure, unless an undue burden would be imposed on the department or agency, that the electronic and information technology allows, regardless of the type of medium of the technology —

(i) individuals with disabilities who are Federal employees to have access to and use of information and data that is comparable to the access to and use of the information and data by Federal employees who are not individuals with disabilities; and

(ii) individuals with disabilities who are members of the public seeking information or services from a Federal department or agency to have access to and use of information and data that is comparable to the access to and use of the information and data by such members of the public who are not individuals with disabilities.

(B) Alternative means efforts: When development, procurement, maintenance, or use of electronic and information technology that meets the standards published by the Access Board under paragraph (2) would impose an undue burden, the Federal department or agency shall provide individuals with disabilities covered by paragraph (1) with the information and data involved by an alternative means of access that allows the individual to use the information and data.

Section 508 requires federal agencies that develop, procure, maintain, or use electronic and information technology to assume responsibility for ensuring accessibility to that technology and its information in a manner comparable to access by federal employees who are not disabled.

Section 508 also requires that people with disabilities who are members of the general public and who seek information or services from the federal government should also have access to that information comparable to people without disabilities.

There are two specific responsibilities here:

- The federal agency is responsible for ensuring that the procurement process includes the accessibility needs of employees with disabilities.
- When involved, industry is responsible for delivering an accessible solution.

In the past, neither entity has taken their responsibility seriously. Federal agencies failed to enforce the standard and industry failed to respond. In 1998, the Access Board recognized this and was again summoned to organize an advisory committee to develop guidelines that ensure the effectiveness of this legislative order. As a member of that advisory committee, I can tell you that one of the key areas we are concerned with involves access to information technologies, including the Internet and the World Wide Web.

Our charter is to assist the Access Board in developing a proposed rule on accessibility standards for electronic and information technology, and it is clearly expected to encompass Web accessibility. It is also expected that this definition of electronic and information technology be consistent with the definition in section 5002(3) of the Clinger-Cohen Act of 1996. (The Clinger-Cohen Act is now the combined Information Technology Management Reform Act and Federal Acquisition Reform Act.)

In addition, the committee will define a broader scope to include future technologies. Then, in conjunction with that new working definition, the advisory group will establish guidelines and standards supporting the definition. As a final process, the Access Board will review the guidelines and subsequently develop the formal federal standard. The committee's work was completed in May 1999. In the spring of 2000, the Access Board and General Services Administration (GSA) released the recommended 508 proposal to the public for comment. Comments were submitted by the end of May 2000 and the mandate is expected to be enacted by the Fall of 2000.

The process for maintaining the effectiveness of this law is reviewed every two years as mandated by the Act. As an additional measure, the Workforce Investment Act of 1998 (http://usworkforce.org/) includes a provision stating that the U.S. Office of Management and Budget establish procedures that require each federal agency to provide written certification indicating that it (the agency) is in compliance with accessibility guidelines. This is to be done annually.

Additional U.S. Policies and Standards That Promote Accessibility

The role of the United States federal government in the establishment of equal rights for all people, including people with disabilities, cannot be underemphasized. This includes legislation, mandates, and legal policy that discuss access to information. Note how the following mandates encourage and, in some cases, specifically enforce creation of accessible information.

The Telecommunications Act of 1996

The Telecommunications Act of 1996 established important standards related to the transmission of information involving telecommunication interfaces and their operating environments. Due to technology convergence, it is very likely that this will include the Internet and World Wide Web.

The Telecommunications Act recognizes the user requirements for people with disabilities and subsequently mandates accessibility. Clearly stating this, Title 1 (Telecommunication Services), Subtitle A (Telecommunications Services), Section 251 (Interconnection) under the topic of "General Duty of Telecommunications Carriers," the Act states: "Each telecommunications carrier has the duty . . . (B) not to install network features, functions, or capabilities that do not comply with the guidelines and standards established pursuant to section 255 or 256."

Wondering what Section 255 and 256 say? The full text of Section 255 appears in the sidebar. For our purposes, the title of Section 255 should suffice: "Access by Persons with Disabilities." Section 255 requires that telecommunications manufacturers and service providers ensure accessibility of equipment and services to people with disabilities. Additionally, to ensure follow through, Section 256 requires the Federal Communications Commission (FCC) to oversee the effort to guarantee interconnectivity of telecommunications networks to people with disabilities.

Section 255 — The Text

Section 255. Access by Persons with Disabilities.

(a) Definitions — As used in this section —

(1) Disability — The term "disability" has the meaning given to it by section 3(2)(A) of the Americans with Disabilities Act of 1990 (42 U.S.C. 12102(2)(A)).

(2) Readily achievable — The term "readily achievable" has the meaning given to it by section 301(9) of that Act (42 U.S.C. 12181(9)).

(b) Manufacturing — A manufacturer of telecommunications equipment or customer premises equipment shall ensure that the equipment is designed, developed, and fabricated to be accessible to and usable by individuals with disabilities, if readily achievable.

(c) Telecommunications Services — A provider of telecommunications service shall ensure that the service is accessible to and usable by individuals with disabilities, if readily achievable.

(d) Compatibility — Whenever the requirements of subsections (b) and (c) are not readily achievable, such a manufacturer or provider shall ensure that the equipment or service is compatible with existing peripheral devices or specialized customer premises equipment commonly used by individuals with disabilities to achieve access, if readily achievable.

(e) Guidelines — Within 18 months after the date of enactment of the Telecommunications Act of 1996, the Architectural and Transportation Barriers Compliance Board shall develop guidelines for accessibility of telecommunications equipment and customer premises equipment in conjunction with the Commission. The Board shall review and update the guidelines periodically.

No Additional Private Rights Authorized — Nothing in this section shall be construed to authorize any private right of action to enforce any requirement of this section or any regulation thereunder. The Commission shall have exclusive jurisdiction with respect to any complaint under this section.

As part of their responsibility, the FCC then requested that the U.S. Architectural and Transportation Barriers Compliance Board (commonly referred to as the Access Board) organize and establish an advisory committee "to assist the Board in fulfilling its mandate under the Communications Act of 1934 as amended by the Telecommunications Act of 1996, Section 255 (hereinafter referred to simply as section 255). Section 255 requires that the Access Board, in conjunction with the Federal Communications Commission (FCC or Commission), develop guidelines, by August 8, 1997, for access to telecommunications equipment and customer premises equipment (CPE) by individuals with disabilities."

Subsequently, the Access Board organized and appointed members to the Telecommunications Access Advisory Committee (TAAC). This committee generated a report containing recommended guidelines to ensure accessibility of telecommunications and customer premises equipment for people with disabilities.

I

2

Cross-Reference

You can find the report "Access to Telecommunications Equipment and Customer Premises Equipment by Individuals with Disabilities" at http://www.access-board.gov/pubs/taacrpt.htm.

The telecommunications industry is taking the Telecommunications Act quite seriously. Understanding the impact of the Act on their constituents, organizations including the Telecommunications Industry Association (TIA) and the Association for Access Engineering Specialists (AAES) (http://www.aaes.org/) — which is jointly operated by the National Association of Radio and Telecommunications Engineers (NARTE) and the Rehabilitation Engineering and Assistive Technology Society of North America (RESNA) (http://www.resna.org/) — are developing new programs to build accessibility awareness for industry members. To address the needs of consumers with disabilities, these organizations are focusing on education and outreach initiatives as well as the implementation of customer service programs that strictly address the needs of the disabled.

To further emphasize accessibility of technology for people with disabilities, the United States government has approved several other measures of legislation. These are briefly described in the next section.

The Assistive Technology Act of 1998

The Assistive Technology Act of 1998 (http://www.itpolicy.gsa.gov/cita/AT1998.htm), or Tech Act, has been instrumental in advancing the needs of individuals with disabilities by providing federal funding to individual states who then use the funding to build awareness about assistive technology. The Act provides grants to assist states in the development of new and innovative programs that assist people with disabilities in the purchase of assistive technology devices and services.

The Tech Act will play a crucial role in the accessibility of information technologies such as the Internet and Web, as shown in conclusions the U.S.

Congress came to as a result of research made to support the legislation. Three of the conclusions noted are as follows:

- Disability is a natural part of the human experience and in no way diminishes the right of individuals to enjoy full inclusion and integration in the economic, political, social, cultural, and educational mainstream of American society.

- During the past decade, there have been major advances in modern technology. Technology is now a powerful force in the lives of all residents of the United States. Technology can provide important tools for making the performance of tasks quicker and easier.

- Many individuals with disabilities cannot access existing telecommunications and information technologies and are at risk of not being able to access developing technologies. The failure of Federal and State governments, hardware manufacturers, software designers, information systems managers, and telecommunications service providers to account for the specific needs of individuals with disabilities results in the exclusion of such individuals from the use of telecommunications and information technologies and results in unnecessary costs associated with the retrofitting of devices and product systems.

These conclusions identify three key motivations for extending the kind of funding Tech Act programs typically enjoy. First, people with disabilities have the right to *full inclusion* of every aspect of life. Therefore, as the Web becomes more of an integral part of American society, the disabled can rightly expect to enjoy the same benefits enjoyed by individuals who are not disabled. Second, technology in general is cited as an important force in the United States, the Web notwithstanding.

The third point recognizes that many people with disabilities do not have access to telecommunications and information technologies (that is, the Internet and Web). However, the most crucial aspect of this finding involves the root cause of why people with disabilities do not have easy or equal access to information technologies. Specifically, the "failure of Federal and State governments, hardware manufacturers, software designers, information systems managers, and telecommunications service providers to account for the specific needs of individuals with disabilities results in the exclusion of such individuals. . . ."

It's sad but very true — information system designers and developers do not include people with disabilities in their standard user interface design or testing. It is precisely for this reason that the World Wide Web eventually evolved into a universal accessibility barrier.

I

2

Thankfully, many of the critical barriers have been removed thanks to efforts that resulted from the Web Accessibility Initiative. Industry, too, is beginning to respond and provide innovative solutions to accommodate people with disabilities in the area of advanced information systems.

Still, while several accessible solutions exist, many of the adjustments are assistive and/or adaptive in nature. Ideally, they should be intrinsic parts of the interface. Good usability design — accessible design — requires user participation by people with disabilities. When this level of participation becomes a standard practice, then and only then will interfaces (regardless of what they are) be truly accessible to people with disabilities.

Legislation such as the Tech Act is designed to promote innovative solutions for advanced information and emerging technologies. A major part of this effort requires the building of awareness of accessibility issues and requirements. The Web has proved to be the most effective and expeditious way of getting the message out and disseminating technology transfer between the disabilities community and high-tech solutions providers.

Legislation in the United States is encouraging accessibility and influencing the development of the World Wide Web. But the United States is just one player in a large pool of participants. Legislation all over the world is addressing Web accessibility, though enforcement power varies.

International Policies and Standards

The importance of international support for an accessible Web cannot be understated. While it's true that many technological and commercial advancements involving the World Wide Web and Internet originated in the United States, it's also true that the Web is playing a key role throughout the rest of the world. This is especially true where access to information technologies for persons with disabilities is concerned.

The effort to increase the accessibility of the World Wide Web is widespread. Nations everywhere are developing policies and standards to enhance the accessibility of electronics and information technology.

The Web Accessibility Initiative itself enjoys a very large contingency of international participants and supporters. The European Community, through its TIDE program, is a founding contributor to the WAI, providing financial support to the initiative. Individual and organizational contributors come from Canada, Australia, Japan, and the United Kingdom. Precisely because of efforts initiated by the European Community, all of Europe is experiencing increased exposure to elements of Web accessibility. Belgium, France, Greece, Norway, Portugal, Sweden, and the United Kingdom are just a few of the countries immersed in Web accessibility activities.

The Web Accessibility Initiative is encouraging this focus through their outreach activities, which have been conducted in more than a dozen countries in Europe alone. This has encouraged an increased level of European participation in the WAI as well as served as a basis for organizing national initiatives to promote Web accessibility.

Notably, official legislation is being enacted in continents and countries including Australia and Canada. In Thailand and Japan, efforts are underway in creating programs or funding initiatives that surrounding nations are patterning their own initiatives after.

Following is an overview of various international policies established to support Web accessibility. Please note that their order is alphabetical and in no way reflects the efforts of one nation over another. Additionally, this is not an exhaustive list — it simply highlights many visible efforts established on an international basis. Guidelines resulting from many of these legal entities are discussed in Chapter 3.

Australia

Right from the start, Australia established itself as a leader in promoting access to the Web for people with disabilities. Australia's Human Rights and Equal Opportunity Commission (http://www.hreoc.gov.au), is the government agency creating and establishing legal decrees mandating equality of access for the disabled.

Key Australian legal standards include the Disability Discrimination Act (DDA) and the Anti-Discrimination Act of New South Wales. Both are discussed in the sections that follow.

Disability Discrimination Act of 1992

The primary legal edict mandating accessibility in Australia is the Disability Discrimination Act (DDA) of 1992. This order is responsible for influencing many of the standards and guidelines that are crucial to defining accessibility throughout the continent. By virtue of the DDA, accessibility measures involving transportation, employment, and insurance have been established. The DDA includes a proviso that allows the Attorney General to make standards and specify rights involving access for people with disabilities. The Attorney General can do this above and beyond what is already stated in the DDA.

As a complement, the Human Rights Commission can advise the Attorney General regarding new or established standards. The Commission has also issued a revised advisory note on accessibility of Web titled "World Wide Web Access: Disability Discrimination Act Advisory Notes."

Australia's Web Accessibility Advisory Note

People who provide goods and services over the Internet need to think about how they can make their WWW sites accessible to people with disabilities. Access can be readily achieved if good design practices are followed.

The Australian Human Rights and Equal Opportunity Commission (HREOC) is drawing attention to resources that will help authors and designers make their World Wide Web documents accessible to the broadest possible audience. In these Advisory Notes, HREOC aims to provide advice about how people can avoid disability discrimination without sacrificing the richness and variety of communication offered by the WWW. These Advisory Notes cannot be exhaustive. In considering any complaints about access, the Commission would take into account the extent to which a service provider has attempted to utilize the best current information and advice wherever it can be found. The Commission does not at this stage believe that there is a single standard against which accessibility can be measured. We do hope that the developments described in these Advisory Notes lead to the emergence of consistent universal guidelines in due course. The AusInfo Guidelines for Commonwealth Information Published in Electronic Formats were launched by the Australian government on 23 March 1999. They are an excellent source of advice about preparing electronic publications.

Few areas compare with the Internet and WWW for pace of change in technical standards. These Notes are not about the content of web pages and have nothing to say about what is appropriate subject matter for publication on the Internet. Nor are the Notes legal requirements. They are advice about good design practices that will help make Web pages, no matter what their content, available to the widest range of people.

Accessibility does not involve an assumption that all pages can be limited to plain text. More sophisticated and innovative pages can and should also be made accessible. In general, this involves provision of alternatives to an otherwise inaccessible feature, rather than any requirement to avoid innovative design.

The rest of these advisory notes provide background information on accessibility and legal issues. As always, comments and suggestions for improvement of these Advisory Notes are welcome.

Of particular note are the guidelines created by AusInfo, an Australian publishing service. AusInfo published a draft of guidelines (discussed in Chapter 3) that addresses the information accessibility needs of people with disabilities.

New South Wales Anti-Discrimination Act 1977

The Australian state of New South Wales (NSW) advocacy level in behalf of people with disabilities is well known and quite admirable. An instrumental piece of legislation released by the New South Wales government that ultimately lead to the creation of Web accessibility is the Anti-Discrimination Act of 1977 (http://www.austlii.edu.au/au/legis/nsw/consol_act/aa1977204/). Part 4A, "Discrimination on the Ground of Disability," deals directly with the levels of accommodation enforced by the Act.

The NSW Attorney General's Department's Guidelines for Web Accessibility, (http://www.lawlink.nsw.gov.au/lawlink.nsf/pages/access_guidelines) now comprise the set of guidelines and standards applicable to government agencies. This not only emphasized the government's effort to create Web accessibility guidelines, but also established the government's lead role in the promotion of Web accessibility.

Canada

Similar to efforts in the United States, Canada is paving the road to ensure accessibility of the Web. Organizations including Industry Canada, the Canadian Human Rights Commission, and Treasury Board Secretariat have worked overtime to create, disseminate, and support legislative efforts including The Canadian Human Rights Act and The Employment Equity Act.

In 1998, the Treasury Board Secretariat's Internet Advisory Committee created a Common Look and Feel working group. This group has the charter of establishing standards and guidelines that will form the basis for official government policy to be adopted and promulgated by the Canadian government.

Another government agency in Canada, the Equity and Diversity Directorate of the Public Service Commission of Canada (PSC), was the first government institution in any country to create a series of Web accessibility guidelines used to evaluate Web pages. Chapter 3 provides additional information about PSC and these guidelines, as well as news regarding Canadian government standards and guidelines.

Not only does this highlight the effect that the Web Accessibility Initiative is having on an international scale, but it stresses how important access to the Web is to government authorities. In Canada, the first goal is to ensure that all government Web sites and associated electronic data is accessible to everyone.

Chuck Letourneau, a former employee of the Canadian government and now a chief consultant and internationally recognized Web accessibility advocate, provides one of the most informative Web accessibility Web sites through Starling Access Services (www.starlingweb.com).

Portugal

In Portugal, efforts on behalf of Internet and Web accessibility have been led by the Portuguese Accessibility Special Interest Group (PASIG). PASIG is a nonprofit organization dedicated to promoting Portuguese initiatives in the accessibility field. While PASIG is not a government agency, their work specifically involves the accessibility of government online information through its proposed Internet Accessibility Guidelines, which are discussed further in Chapter 3.

Worthy of note is the establishment of an International Accessibility Board consisting of well-known accessibility experts and industries. Accessibility Board members in collaboration with PASIG helped to compile the Internet Accessibility Guidelines submitted to the Portuguese Parliament on February 17, 1999 for acceptance and adoption.

Portuguese International Accessibility Board

Name	Affiliation
David Bolnick	Microsoft Corporation
Geoff Freed	WGBH/NCAM
Phill Jenkins	IBM
Earl Johnson	Sun Microsystems
Neill McBride	Dolphin Computer Access
Emmanuelle Gutiérrez y Restrepo	SID@R
Javier Romañach	COCEMFE
Cynthia D. Waddell	ADA Coordinator, City of San Jose, CA

United Kingdom

The Disability Discrimination Act (DDA) is the primary piece of legislation guaranteeing rights to people with disabilities in the United Kingdom. Much of the DDA is modeled after the Americans with Disabilities Act, using the term *reasonable adjustment* in place of the ADA's *reasonable accommodation*.

Similar to the ADA, the Disability Discrimination Act establishes and defines a series of rights for disabled people involving employment, provision of goods, facilities, and services. DDA places responsibility upon employers and suppliers of goods and services to provide reasonable adjustment and, more importantly, to remove any barriers that could result in discrimination.

The Disability Net Web site (`http://www.disabilitynet.co.uk/info/legislation/ddaguide/legislation.html#TDDA`) includes a lengthy discussion and interpretation of the UK's Disability Discrimination Act that you will find useful.

Summary

This chapter introduced you to legislation and international law promoting Web accessibility and general rights for people with disabilities. You discovered that the effort to increase access for people with disabilities is moving swiftly throughout the world and is not just a United States thing. In addition, you learned the following:

- Individuals with disabilities are using the Americans with Disabilities Act (ADA) to exercise their rights to information on the Web.

- The National Information Infrastructure (NII) has established goals and guidelines to guarantee access for all people, including those with disabilities.

- The U.S. Department of Justice has determined that the Americans with Disabilities Act guarantees effective and accessible communication, including access to information, programs, services, and goods made available through the Internet and World Wide Web.

- Section 508 of the Rehabilitation Act, the Assistive Technology Act, and the Telecommunications Act are key U.S. policies establishing access to information technologies for the future.

- Nations and continents including Australia, Canada, and the United States are approving specific legislation to enforce Web accessibility.

- Additional funded efforts and policies in the European Community and Asia demonstrate growing international support for Web accessibility.

I

2

References

"National Federation of the Blind Sues America Online, Inc." National Federation of the Blind Press Release, November 1999. Available from http://www.nfb.org/AOLpr.htm.

"Access to Telecommunications Equipment and Customer Premises Equipment by Individuals with Disabilities." Telecommunications Access Advisory Board Final Report, January 1997.

United States Department of Justice, Civil Rights Division, Freedom of Information Act, Americans with Disabilities Act Core Letters. *Accessibility of Web Pages on the Internet.* Assistant Attorney General, Deval L. Patrick, Core Letter #204, response to Hon. Senator Tom Harkin, 1996.

Anti-Discrimination Act of 1977, New South Wales. Available from http://www.austlii.edu.au/au/legis/nsw/consol_act/aa1977204/.

Waddell, Cynthia. "Applying the ADA to the Internet: A Web Accessibility Standard." Paper presented at the American Bar Association National Conference, 17 June 1998. Available from http://www.isc.rit.edu/~easi/law/weblaw1.htm.

Australia Disability Discrimination Act of 1992. Available from http://www.austlii.edu.au/au/legis/cth/consol_act/dda1992264/.

The Canadian Human Rights Act, 1998.

National Information Infrastructure Advisory Council. *Common Ground: Fundamental Principles for the National Information Infrastructure,* March 1995.

Raspberry, William. "Complaints Against Common Sense." *Washington Post,* 16 November 1998.

The Disability Discrimination Act of 1995. Available from http://www.disability.gov.uk/dda/.

Beer, Matt. "Disabled Issues Confront Web Sites." *San Francisco Examiner,* 12 November 1998.

The Information Technology Management Reform Act of 1996. Available from http://www.rdc.noaa.gov/~irm/div-e.htm..

The National Disability Law Reporter. Volume 10, Issue 6, 1997.

United States Department of Education, Office of Civil Rights. Settlement Letter, Docket Number 09-95-2206. 1996.

United States Department of Education, Office of Civil Rights. Settlement Letter, Docket Number 09-97-2002. 1997.

Tyler v. City of Manhattan, 857 F. Supp. 800 (D.Kan. 1994).

WAI's Policy References. Available from `http://www.w3.org/WAI/References/Policy`.

Workforce Investment Act. *Federal Register,* April 1999. Available from `http://usworkforce.org/finalregs.txt`.

Gore, Al. Speech presented at the World Telecommunication Development Conference. Buenos Aires, 21 March 1994.

"World Wide Web Access — Disability Discrimination Act Advisory Notes." Disability Rights Unit, Human Rights and Equal Opportunity Commission, March 1999. Available from `http://www.hreoc.gov.au/disability_rights/standards/www_3/www_3.html`.

3

Chapter 3

Using Standards and Guidelines

In This Chapter

- The WAI guidelines
- Standards
- Best practices and guidelines
- Industry guidelines
- Additional support guidelines

The creation of written guidelines and standards addressing Web and Internet accessibility is a key reason why the World Wide Web is on its way to becoming more accessible.

Chapter 2 introduced you to the legal parameters enforcing Web accessibility. This chapter focuses on the standards, guidelines, and best practice documents that establish the criteria and requirements for building accessible Web sites.

As you examine the guidelines, try to look for common threads — many exist among them. For the most part, existing standards and Web accessibility guidelines rely on a consistent set of user characteristics and requirements. Note also that several written documents are "works in progress." Between technology advances and evolving user preferences, it is necessary to maintain an open dialog allowing for change.

Note

Every recommendation and guideline document cited in this chapter is available on the Web in electronic form. If your employer or organization is considering the development of internal standards, be sure to download the documents for reference.

The WAI Guidelines

The Web Accessibility Initiative or WAI program office is responsible for the World Wide Web Consortium's commitment to Web accessibility for all people. The office is intensely involved with the creation of industry supported guidelines for content design, publishing tools, and user agents.

Currently, the WAI has three primary working drafts:

1. Web Content Accessibility Guidelines
2. Authoring Tool Guidelines
3. User Agent Guidelines

Each of these documents is reviewed in the sections that follow.

Web Content Accessibility Guidelines

The official name of this set of guidelines is "W3C Web Content Accessibility Guidelines 1.0." The guidelines are a W3C-proposed recommendation and a work in progress. You can download them from the World Wide Web Consortium's Web site at `http://web1.w3.org/TR/WAI-WEBCONTENT/`.

I

The Web Content Accessibility Guidelines are *the* authority for designing and creating accessible Web sites.

Web Sites Promoting Guidelines

The W3C's Web Accessibility Initiative (WAI) is the recognized authority and central repository for the development and promotion of Web accessibility guidelines. However, there are other organizations and Web sites that you should take note of as you endeavor to build an accessible Web site. These include the following:

- IBM Guidelines for Writing Accessible Applications Using 100% Pure Java
  `http://www-3.ibm.com/able/snsjavag.html`
- Microsoft's Guidelines for Accessible Web Pages
  `http://www.microsoft.com/enable/dev/web/guidelines.htm`
- Oregon State University Web Accessibility Guidelines
  `http://tap.orst.edu/Policy/web.html`.
- Santa Rosa Junior College Web Accessibility Checklist
  `http://www.santarosa.edu/access/www/checklist/`.

Recognized experts in the area of accessibility served as authors and reviewers of the documents. Keep these in a high-profile location in your browser's bookmarks or favorites section, pointing to the URL the W3C always uses to store the most recent version of the document in question.

The purpose of the Web Content Accessibility Guidelines as stated within its abstract is as follows:

> to promote accessibility . . . following them will also make web content more available to *all* users, whatever user agent (for example, desktop browser, voice browser, mobile phone, automobile-based PC, and so forth) or constraints they may be under.

The W3C WAI WCA Working Group recognizes that user agents (for example, Web browsers) are evolving entities.

We started out with text browsers such as Lynx, moved to a GUI (graphical user interface) interface with Mosaic, and are now at the point where anything that emits sound or includes a display can render Web content. As a result, designing content must be done in a way that enables flexible rendering. This is not just important for an individual with a disability, but also for devices and interfaces that are considered common user interfaces — televisions and cell phones, for example.

The Web Content Accessibility Guidelines also remind us of another important point: Accessibility does not mean minimal page design.

Often, Web page developers/designers who are not familiar with users with disabilities believe that Web accessibility specialists recommend the development of Web sites that are vanilla, bland, and text-based. That is, no pictures, no color, no flair or style, no programming, and by all means, nothing that reaches out and grabs you. This simply is not true.

Let's assume that your Web site is a business based on the movie industry. You sell and distribute videos, provide reviews, and perhaps include recorded interviews of actors and actresses. Quite obviously, developing a Web site involving movies should include video clips. With a little forethought and entrepreneurial spirit, you can set up Web servers that generate full motion pictures. To increase the accessibility of your site, all you need to do is ensure that the movies are served with captioning and audio description features. Today, Web technology that embeds both of these accessible features is available (for example, MagPie, SMIL, SAMI, and QuickTime). By doing this, you'll not only increase your site's accessibility level, but you'll also expand your customer base to include blind and deaf people.

Tip

WGBH Educational Foundation in Boston, home to the Descriptive Video Service, currently has a collection of more than 200 titles including descriptive videos. Hundreds of other titles include captioning for the deaf. If you need assistance creating either descriptive video or captioned versions of your video, refer to WGBH's National Center for Accessible Media (NCAM) on the Web at http://www.wgbh.org/ncam. Additionally, NCAM recently released an new tool that makes it easier to created captioned audio and video for the Web called MAGpie (Media Access Generator). You can download MagPie for free from the NCAM homepage.

The goal of the guidelines is to encourage good design practice. The focus is to promote the design of Web sites that are highly usable for the greatest number of surfers. Your Web site's intrinsic value is the information you create for it. Accessible page authoring enables you to distribute that information to a broader audience. Keep in mind that these guidelines were created to serve as an exhaustive resource – reading the guidelines can be tedious. The dividends are well worth the effort!

The W3C Web Content Accessibility Guidelines are in-depth and detailed. A helpful feature of the guidelines is that they are prioritized and come with an associated checklist (`http://www.w3.org/TR/WAI-WEBCONTENT/full-checklist`). Additionally, the Web Content Accessibility Guidelines Working Group created a Techniques document that assists you by providing examples of code and user scenarios. You can access this document at `http://www.w3.org/TR/WAI-WEBCONTENT-TECHS`.

Following is a list of Content Accessibility Guidelines designated as high priority items. Topics include Web page design involving images, programming scripts, navigation, multimedia, forms, frames, and tables. Part II of this book focuses on how to implement many of these guidelines. This subset of the guidelines was extracted directly from the current working draft:

1. Provide equivalent alternatives to auditory and visual content.
2. Don't rely on color alone.
3. Use markup and style sheets properly.
4. Clarify natural language use.
5. Create tables that transform gracefully.
6. Ensure that pages featuring new technologies transform gracefully.
7. Ensure user control of time-sensitive content changes.
8. Ensure direct accessibility of embedded user interfaces.
9. Design for device-independence.
10. Use interim solutions.
11. Use W3C technologies and guidelines.
12. Provide context and orientation information.
13. Provide clear navigation mechanisms.
14. Ensure that documents are clear and simple.

Authoring Tool Accessibility Guidelines

The one user request that may outnumber requests for accessible Web pages is the need for authoring tools to support accessibility features. Trying to remember a set of guidelines can be a difficult task. An authoring tool that automates the accessibility process becomes a value-added productivity aid. This is principally true for Web page authors who are interested in accessibility, but simply don't have the time to become familiar with or keep up on all the developments.

Until recently, very few authoring tools supported Web accessibility features. At best, some tools include prompting for alternative text (for example, **alt = " "**). Others support certain HTML elements or attributes that increase accessibility. Only one authoring tool, HoTMetaL PRO, implements accessibility prompting and validation. (See Chapter 8 for more information about HoTMetaL PRO's accessibility features.)

The W3C WAI program office tackled the situation by instituting the Authoring Tool Guidelines Working Group whose charter is to create a set of guidelines for developers of Web publishing tools. These guidelines provide recommendations for implementing support features or mechanisms within the publishing applications. By encouraging the development of publishing tools that support the generation of accessible content, the WAI program office is raising the percentage of accessible Web sites by default. Web site designers are much more likely to implement accessible Web code when their support tools prompt, validate, and automatically repair Web pages for accessibility.

The complete working draft of the Authoring Tool Accessibility Guidelines can be found at `http://www.w3.org/WAI/AU/WAI-AUTOOLS-19990621/wai-autools.html`. Following are a few of the working draft guidelines that will help form the first official recommendation:

- Support all accessible content recommendations
- Generate standard markup
- Identify all inaccessible markup
- Ensure that all markup inserted by the authoring tool is accessible
- Ensure that conversion tools produce and retain accessible markup and content
- Provide comprehensive accessibility help to authors
- Ensure that users may configure accessibility mechanisms

User Agent Accessibility Guidelines

User agents usually refer to the browser or application you use to view a Web page. Netscape Navigator, Microsoft Internet Explorer, Lynx, pwWebSpeak, and Opera are examples of common user agents. WebTV is a user agent. Pagers, cell phones, and kiosks are also types of user agents. In effect, almost any device or technology capable of rendering Web content is a user agent. Accordingly, the accessibility of the user agent is vital to the user.

Ideally, developers of user agents try to design interfaces that can be used by the widest range of users possible. On the other hand, it's perfectly valid to create an application-specific browser. For example, a user agent designed for an automobile is much more practical if the interface is based on speech input/output technology. You would hardly expect (or want) the driver to be put in position where keyboard typing or mouse movement was required to operate the vehicle. On the other hand, an interface that enables the driver to voice commands and then generates an audible answer is safe way to drive and surf.

The WAI User Agent Accessibility Guidelines provide solid direction regarding the development of accessible user agents. Please refer to the full text of these guidelines at `http://www.w3.org/TR/WD-WAI-USERAGENT/`.

Again, please note that the User Agent Guidelines have been designated by the W3C as a working draft and are subsequently open to future revisions. What follows is a useful subset of the proposed guidelines. Developers will find these helpful for designing accessible user agents.

- Ensure that the user interface is accessible.
- Ensure that user agent accessibility features are configurable.
- Ensure that users can disable features that might interfere with accessibility.
- Provide summary information about keyboard access.
- The user agent must render information accessibly.
- Allow the user to control document styles.
- Provide access to alternative representations of content and control of its rendering.
- Provide information about the content and structure of a document and the user interface.
- Provide information about events that occur.
- Allow keyboard navigation of the document and views of the document.
- The user agent must make information available to other technologies.
- Use and provide accessible interfaces to other technologies.

Standards

The most effective way to ensure that accessibility becomes an integral part of industry and technology — any technology — is to define what constitutes accessibility in a standard. Standards are requirements subjected to attainable levels of quality. In the past, information technology standards organizations and working groups rarely considered the accessibility of technologies important enough to include in the standards process. This was based on an unfounded perception that accessibility standards were not useful to the "average" person and subsequently presented too difficult a challenge for the developer to implement.

Recently, that mindset has changed. A search on standards documents through the American National Standards Institute's NSSN: A National Resource for Global Standards (http://www.nssn.org/) returned no less than 25 standards documents (international and national) that included specifications and considerations for people with disabilities.

Consider the information in the following sections regarding international standards.

International Standards

The International Organization for Standardization (ISO) Software Ergonomics Working Group, *TC159/SC4/WG5,* whose primary responsibility involves multimedia, has taken on the additional assignment of creating an accessibility standards document. According to a report distributed by Harry E. Blanchard, of the American Computing Machinery (ACM) organization, the Working Group is planning to use the ANSI/HFES-200 standard as a base for creating the ISO standard. Note that at the time this writing, no additional information regarding the standard could be found.

The International Electrotechnical Commission (IEC) is working on a multimedia standard, the *IEC TC100,* which contains a provision for disabilities access. Work on this standard is just starting, and a first draft is expected early in 1999. In the past, IEC has developed standards for disabilities equipment, including hearing aids.

At this point, then, it is fair to say that international standards organizations are still well behind in the creation of standards that promote accessible design. However, with the growing support for accessibility by smaller government, industry, and national standards organizations, it's reasonable to expect that mindsets will change. Many of these standards are identified and discussed in the next few sections.

National

Work on the national standards front is making great progress. These standards are likely to be very influential on the future design of the World Wide Web. An example of this is the work accomplished by the Human Factors and Ergonomics Society (HFES) (http://www.hfes.org/) and American

National Standards Institute (ANSI) (http://web.ansi.org/) involving the ANSI/HFES-200 standard. This standard specifically addresses user interface and general software requirements involving people with disabilities.

ANSI/HFES-200 Software User Interface Standard

The objective of the ANSI/HFES-200 standard is to provide design requirements and recommendations to increase the effective usability of software interfaces. The ANSI/HFES-200 Software User Interface standard includes a specific entry (Part 2) for accessibility.

Section 5.3 of the ANSI/HFES-200 standard notes the following regarding Part 2 of the standard regarding accessibility:

> The recommendations in HFES-200 Part 2 focus on features and functions of computer operating systems, drivers, application services, other software layers upon which applications depend, and applications that increase the accessibility of applications for users with disabilities. Hardware is not specifically addressed by any recommendations; however various hardware assistive devices may exploit recommended functions that are provided by operations system and application software. Accessibility of applications is often facilitated by a combination of both hardware and software. The guidelines on Accessibility are aimed at reducing the need for add-on assistive technologies (hardware and software) while promoting increased usability of systems using such add-on technologies, when they are required. The proposed standard on Accessibility does not address the behavior or requirements for assistive technologies themselves (including assistive software), but instead focuses on enabling features provided by operating systems, drivers, application services, and other software upon which applications depend.

The scope of ANSI/HFES-200 is on user interaction with software for personal and business use. However, some of the recommendations apply to home and mobile computing, as well as interactive voice systems. Many of the recommendations apply to future design of Web applications.

The ANSI/HFES-200 standard entered its final review during the later half of 1998. It is expected to be available to the public sometime in 1999. This standard specifies accessibility measures for user characteristics, keyboard input, pointing devices, audio output, and general accessibility guidelines.

Best Practices and Guidelines

A product designed accessibly often enhances the usability of that product for all people, regardless of ability. Very often, accessible design features go unnoticed. For example, have you observed how public telephones include volume control buttons? Most likely, you've used the volume control feature when you've called someone from a noisy airport or from a street phone. This feature is so common you might forget that volume control is an important accessibility feature created to assist a person with a hearing impairment.

Your Web browser includes accessible features too. If you use Microsoft Internet Explorer (shown in Figure 3.1) or Netscape Navigator (displayed in Figure 3.2), you'll note that both contain menu items that enable you to increase the font size of a Web page so you can read or view the content with better visual clarity. Many people use these accessibility characteristics — not just people with visual impairments.

Figure 3.1 **Internet Explorer increases or decreases font size with the fonts menu item**

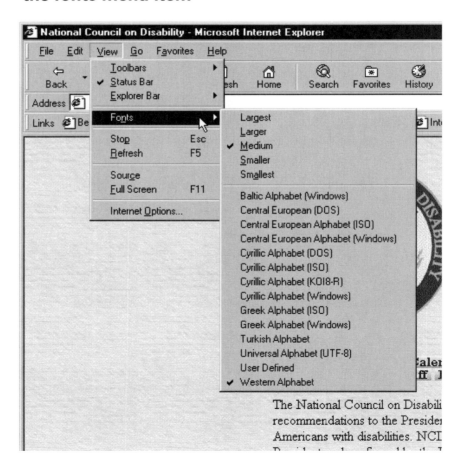

**Figure 3.2   Netscape Navigator uses Increase Font/Decrease Font
to adjust character size**

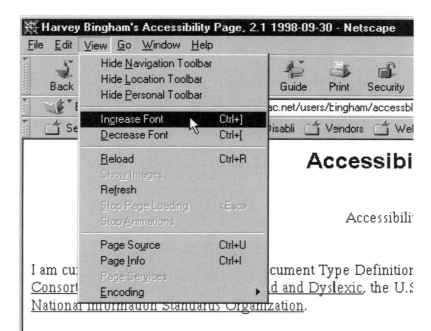

Another good way to promote accessible design of the Web is by developing guidelines or best practice documents. Over the past two or three years, this strategy has resulted in several publicly available documents. The following sections contain descriptions of these documents and highlight their useful features. (Note the international availability of the documents.)

IEEE Best Practices

The Institute of Electrical and Electronics Engineers (IEEE) Computer Society sponsors the Internet Best Practices Working Group (IBPWG), which is chartered to create the IEEE Std. 2001-1999 best practices document.

The purpose of the standard is to develop a series of "recommended practices and requirements set forth . . . to reduce (but not eliminate) the risks associated with Web page investments." The standard's focus is to establish guidelines for creating "well engineered pages."

Jim Isaak, Working Group Chair, notes that the standard emphasizes that "Web site design is an application engineering task . . . It starts with requirements and design . . . and continues from implementation to maintenance and end-of-life processes." Jim also points out that the standard highlights how "accessibility needs to be understood in terms of design and requirements, and then incorporated into all follow-on aspects of the site life cycle."

The Best Practices standard cross-references the W3C WAI guidelines as a key resource for designing accessible Web sites.

The HARMONY Guidelines

In 1994, the Commission of the European Union launched the HARMONY Project under its Technology Initiative for Disabled and Elderly People (TIDE) program. The objective of the project was to increase the availability and quality of print information for print disabled people. HARMONY concluded in December 1996 with the release of a final report that includes guidelines involving good and bad practices for text anchors, images, line graphics, audio clips, image maps, forms, and tables. The document also provides instructions for testing and embedding descriptive text links. You can download the complete HARMONY document at `http:// www.esat.kuleuven.ac.be/teo/docarch/projecten/harmony/ harmony.en.htm`.

I found the HARMONY good practices (Annex A) very interesting and easy to understand. Several principles are documented in other guidelines, but the clarity of the HARMONY recommendations is worth noting. Here are a few of the guidelines suggested. First, the good practices:

- Maintain a standard and consistent page layout.
- Make hypertext links descriptive enough so that they may make sense when read out of context.
- Provide all graphic and audio information in alternative formats.
- Provide alternatives for all links accessible through an image map, usually by having a list of hypertext links just below the image map.
- Include HTML, or at least ASCII, versions of all documents presented in PDF, Postscript or other formats.

- Tables should have paragraph or line breaks at the end of each cell. Accessibility can also be improved by minimizing the number of columns used, since a more "vertical" layout aids left-to-right and top-to-bottom access.

- In addition to citing good practices, the HARMONY guidelines specify three bad practices or practices to avoid:

- Avoid using HTML constructs (tags) that are non-standard or specific to one Web browser.

- Presentation of text in columns should be avoided as it can cause problems for certain access technologies, especially screen readers (for the blind).

- Do not use graphics to provide organization or structure to the document, use appropriate HTML tags and suitable punctuation.

- As a final reminder, the HARMONY guidelines recommend testing Web pages with a variety of browsers. Annex B of the HARMONY document includes specific Web accessibility guidelines and includes examples of code.

The Nordic Guidelines

In 1998, the Nordic Cooperation on Disability released the second version of the Nordic Guidelines for Computer Accessibility. The Cooperation is directed by the Nordic Council of Ministers and supported by the governments of Denmark, Finland, Iceland, Norway, and Sweden.

The Nordic Guidelines for Computer Accessibility provide accessibility requirements for personal computers in order to encourage access and use by people with disabilities. The documents notes: "The emergence of the World Wide Web has made it possible for individuals with appropriate computer and telecommunications equipment to interact and access information as never before." Citing the importance of the World Wide Web as an information infrastructure, the Nordic Guidelines recommend the following:

- Text alternatives to graphics
- Text-only pages to be updated in parallel to main graphics pages
- Alternative text to image maps
- Use of standard HTML formats and tags

The Nordic Guidelines conclude with a definitive requirement: "Web page designers should be requested to adopt or consult existing guidelines for making accessible Web pages."

The Portuguese Internet Accessibility Guidelines

The Portuguese Accessibility Special Interest Group (PASIG) formed a group of international experts to create a series of accessibility standards for the Web. The PASIC guidelines were originally intended to be applied to online information published by the Portuguese government and public services. PASIG also hopes that companies, organizations, and individual users will use the guidelines.

The PASIG Internet Accessibility Guidelines are divided into mandatory guidelines and recommended guidelines. They cover three primary areas: information presentation, navigation, and compliance. The final version of the draft is available at `http://www.acessibilidade.net/doc/acessibilidade/guidelines_draft2.html`.

As is the case with many guidelines, the Portuguese guidelines are patterned after the Web Accessibility Initiative guidelines.

Canadian Guidelines

In 1998, Canada's Treasury Board Secretariat's Federal Identity Programme created a Common Look and Feel Working Group. This group has the charter of recommending standards and guidelines that will form the basis for official government policy to be adopted and promulgated by the Canadian government.

Chuck Letourneau, A key member of the Secretariat's Internet Advisory Committee, recently reported that, "Among the factors considered by the Common Look and Feel WG was the need to incorporate work already done by others, including the Access Working Group of the Internet Advisory Committee and international organizations (for example, W3C and W3C/Web Accessibility Initiative). With respect to accessibility, the Common Look and Feel WG has recommended that a requirement to comply with all Priority 1 and Priority 2 checkpoints in the WAI Web Content Accessibility guidelines set be included in the policy." Letourneau is hopeful that this will become policy in 1999.

Letourneau then noted, "A sub-group of the Common Look and Feel WG has come up with six quick fixes, one of which is the requirement to provide all images on these primary pages with suitable alt-text."

Not only does this highlight the effect that the Web Accessibility Initiative is having on an international basis, but it stresses how important access to the Web is to government authorities. In Canada, the first goal is to ensure that all government Web sites and associated electronic data is accessible to everyone.

In addition to the work being accomplished by the Treasury Board Secretariat, another government agency in Canada, the Equity and Diversity Directorate of the Public Service Commission of Canada (PSC), has long been a promoter of Web accessibility. In fact, the PSC was the first government institution to create a series of Web accessibility guidelines used to evaluate Web pages. Before Web accessibility validation tools such as Bobby and WebSAT were available, the PSC created a set of simple guidelines along with a self-evaluation tool which served as useful aids in constructing accessible Web pages. The PSC site (http://www.psc-cfp.gc.ca/) also includes a self-evaluation test for using the JavaScript scripting language.

Industry Guidelines

International standards, guidelines, and best practice documents are good ways of increasing access to the Web. However, the most effective means for increasing accessibility occurs when electronic commerce corporations develop, disseminate, and build awareness about their own accessibility standards. This is becoming a standard practice for several high-tech companies that have established themselves as leaders in the drive toward an accessible Web.

The benefits of this kind of business policy are far reaching because the focus is not on profits or sales, but rather on building awareness. Not only do employees and customers with disabilities experience the immediate effects of accessible interfaces, but business partners, channels, value-added resellers (VARs), and original equipment manufacturers (OEMs) also benefit from increased awareness.

As business visionary John Kao aptly stated in his book, *Jamming*: "Without awareness, the organization is flying blind. You need to penetrate a company's boundaries, to challenge its institutional defenses and prejudices . . . Awareness must be systematically cultivated. It often comes from the conceptual and cognitive divergence created when the right mix of people and resources are thrown together."

That "right mix" may never be as evident as it is in the three leading corporations involved in Web accessibility: IBM, Microsoft, and Sun Microsystems. The following sections highlight their standards and guidelines activities.

IBM Web Page and Java Accessibility

IBM Special Needs Systems (SNS) is the "granddaddy" of corporate disability organizations today. Over the years, IBM has developed both hardware and software products to assist people with disabilities. They were responsible for promoting the first set of PC operating system software tools and have been vigorously involved in accessibility standards and guidelines from the start.

IBM SNS has simplified the Web accessibility guidelines process by focusing on the essentials of accessible Web design while still supporting the full version of accessible Web guidelines produced by the WAI. The SNS Web accessibility checklist (`http://www-3.ibm.com/able/accessweb.html`) is similar in content to the WAI Quick Tips Reference Card (`http://www.w3.org/WAI/References/QuickTips`), both of which highlight the following:

1. Use `ALT="TEXT"` attribute to describe the function of visuals.
2. Use client-side image maps.
3. Provide captions or transcripts for audio. Provide text or audio description for video.
4. Provide alternate content for scripts, applets, and plug-ins.
5. Markup content with proper structural elements. Use Cascading Style Sheets (CSS2).
6. Summarize content of each graph and chart.
7. Provide a title for each frame.
8. Use the header, caption, and summary tags for tabular data.
9. Use descriptive link text.
10. Check accessibility using the available tools.

The IBM checklist also recommends that page authors associate labels explicitly with form elements.

In the past year, IBM SNS partnered with Sun Microsystems to design accessibility into the popular Java Web programming language. In conjunction with this work, IBM produced a full set of publicly available Java application development guidelines for accessibility. You can find the complete set of guidelines at `http://www-3.ibm.com/able/accessjava.html`.

Cross-Reference

Chapter 9 discusses the Java application development guidelines and Java accessibility in depth.

IBM SNS recommends the following checklist items to simplify the Java application accessibility programming process:

1. Use the Java Foundation Classes/Swing Set.
2. Follow the essential accessibility practices:
   - Enable keyboard navigation.
   - Describe icons and graphics and set Accessible Description on all components.
   - Set the focus.
   - Label components.
   - Name logical groups.
   - Be sufficiently multithreaded.
   - Provide a logical layout.
3. Follow the guidelines when extending (or not using) JFC/Swing Set.
4. Test for accessibility using assistive technology (self-voicing kit technology for Java).

One of the most challenging and barrier-filled Web applications available today is Lotus Notes. At one point, things were so bad I thought there would be a national uproar by disability organizations across the United States. Shortly after IBM purchased Lotus Corporation, the Special Needs Systems took the proverbial bull by the horns and set out to enhance Notes accessibility. There is still plenty of work to be done, but SNS has constructed a brief list of guidelines for creating accessible Lotus Notes applications. You can review these guidelines at `http://www-3.ibm.com/able/accessnotes.html`.

Lastly, IBM SNS also produced a set of software accessibility guidelines, including principles of accessible software design (`http://www-3.ibm.com/sns/accesssoftware.html`). I encourage you to review these guidelines and distribute them amongst your colleagues.

Microsoft Web Accessibility

In May 1998, I was invited by Microsoft's Accessibility and Disabilities group to participate in the much-heralded "Defend Windows 98 Release" press conference in New York City. It was an interesting experience — one I expect I'll tuck back in my book of memories for years to come. What I found interesting were the questions I was asked by members of the media after the press conference. More specifically, one particular question I encountered several times was: "How much did Microsoft pay you to come here?" The answer: Not one cent. (I suspect that this is one reason why Bill Gates is wealthy and I'm a struggling consultant. He's paid for personal appearances.)

My point is reasonable enough. It doesn't matter to me whether you're Microsoft or John Doe's Cheap Software — if you're creating accessible products and/or building accessibility awareness, I'm in your corner. I wasn't at the press conference to hype Microsoft. I was there to build awareness about the Windows 98 accessibility enhancements, including its built-in screen magnifier for people with low vision.

Microsoft has made tremendous strides in the advancement of accessible software at every level — operating systems, applications and networks, and particularly Web accessibility. Both Windows 2000 and Windows 98 include their own built-in screen magnifier, screen reader and on-screen keyboard! Microsoft's Accessibility and Disabilities Web page is an information metacenter for accessibility awareness. I would highly recommend two areas for developers and software engineers looking for guidelines and Microsoft development standards: `http://www.microsoft.com/enable/dev/default.htm` and `http://www.microsoft.com/enable/products/office2000/default.htm`. Microsoft also provides text versions of these documents.

In addition to operating system and software application guidelines, Microsoft has developed guidelines for Web pages (`http://www.microsoft.com/enable/dev/web/intro.htm`), guidelines for accessible Web pages for Internet Explorer (`http://www.microsoft.com/enable/dev/web/guidelines_IE.htm`), and detailed information about Microsoft Active Accessibility for Java (`http://www.microsoft.com/java/resource/access.htm`).

Like IBM, Microsoft also provides a checklist containing 12 guidelines for creating accessible Web pages:

1. Use good ALT text for all graphics.
2. Use image maps properly.

3. Provide useful link text.
4. Enable good keyboard navigation.
5. Provide alternatives to all controls and applets.
6. Create alternate pages that don't use frames.
7. Ensure proper use of tables and their alternatives.
8. Support the reader's formatting options.
9. Don't require the use of style sheets.
10. Use file formats the reader can use.
11. Avoid using scrolling marquees.
12. Provide titles for most objects.

Microsoft concludes by recommending that you test your pages for accessibility using the Bobby Web accessibility validation service.

Sun Java Accessibility

Sun's Enabling Technologies Program is the key provider of guidelines for Java accessibility, as you might expect. Sun provides an excellent primer (http://www.sun.com/access/) for creating accessible Java applications. Please refer to Chapter 9 for a thorough discussion of Java accessibility.

Additional Support Guidelines

Before the Web Accessibility Initiative (WAI) program office was organized, several disability research and support organizations developed guidelines to encourage the development of accessible Web pages, browsers, tools, and validation services. Many of these guidelines formed the basis for the WAI document titled "W3C Web Content Accessibility Guidelines."

I believe it's important to recognize the organizations involved with Web accessibility and their specific contributions. Note that there are hundreds of organizations worldwide that have contributed in one way or another to the development of Web accessibility guidelines.

Cross-Reference

Most of the following guidelines are described in Chapter 13.

NCSA's Mosaic Access Project

The first organization funded by the National Science Foundation to develop research data regarding people with disabilities and access to the World Wide Web, the National Center for Supercomputing Applications (NCSA), was hosted by the same university responsible for the development of Mosaic, the first graphics-based Web browser.

The Mosaic Access Project first developed an initial set of issues, concerns, and barriers involving access to Web content and development of browsers. This information was organized into an online document you can find at http://bucky.aa.uic.edu/. The project has since been discontinued, though several of the original steering committee members (http://bucky.aa.uic.edu/HTML/map.html) are associated with the W3C's WAI program office.

Trace Research and Development Center

Trace Research and Development Center is chiefly responsible for compiling and publishing the original set of Web accessibility guidelines. Much of what Trace developed has since been absorbed into the WAI guidelines. In fact, one of the primary editors for the W3C Web Content Accessibility Guidelines is Wendy Chisholm, an engineer at Trace.

Trace's Web site is a model of accessibility. You can find all their guidelines and developments related to Web accessibility at http://www.trace.wisc.edu/world/web/.

Adaptive Technology Research Centre (ATRC) at University of Toronto

ATRC is a leader in the development of emerging technology and Web interfaces. ATRC, in conjunction with SoftQuad, assisted in the development of the first publishing tool that supported Web accessibility. HoTMetaL PRO not only holds the distinction of being the first ever graphics-based Web publishing tool, but to this day it is the only publishing tool that implements accessibility validation and prompting. ATRC also maintains a Web site that includes guidelines for Web accessibility development. Along with standard Web site accessibility guidelines, ATRC features Cascading Style Sheet (CSS) and Virtual Reality Modeling Language (VRML) implementations. They have compiled this information in a nicely organized online slide show that you can review at http://www.utoronto.ca/atrc/rd/slideshows/inclusive.html.

WGBH's National Center for Accessible Media (NCAM)

The National Center for Accessible Media (NCAM) has been the catalyst for establishing and promoting guidelines involving Web captioning and audio description that ensures the accessibility of Web-based multimedia. Instructions and guidelines for creating accessible multimedia can be found on NCAM's Web Captioning and Audio Description Web page at `http://www.wgbh.org/wgbh/pages/ncam/webaccess/captionedmovies.html`.

Cross-Reference

Chapter 10 provides specific examples of captioning and audio description.

Summary

In this chapter, you learned about standards, guidelines, and best practice documents that specify Web accessibility requirements. You also learned the following:

- The Web Accessibility Initiative provides guidelines for page authoring, user agents, and authoring tools.
- International standards bodies including ISO and the IEC are beginning to support standards for accessible Web sites and multimedia.
- The objective of the ANSI/HFES-200 standard is to provide design requirements and recommendations to increase the effective usability of software interfaces. The ANSI/HFES-200 standard is likely to be the first official U.S. industry standard supporting Web accessibility.
- Electronic commerce and software vendors are influencing Web accessibility by creating and supporting corporate accessibility standards.
- IBM, Sun Microsystems, and Microsoft have created Java accessibility standards.

References

The HARMONY Consortium. "The Art of Providing Access to Electronic Documents — The HARMONY Final Report," 1997. Available from `http://www.esat.kuleuven.ac.be/teo/docarch/projecten/harmony/harmony.en.htm`.

Equity and Diversity Directorate of the Public Service Commission of Canada Web site. `http://www.psc-cfp.gc.ca/eepmp-pmpee/internet_home.htm`.

IBM Corporation. "IBM Software Accessibility Guidelines." Available from `http://www-3.ibm.com/able/accesssoftware.html`.

IBM Corporation. "The IBM Special Needs Systems Web Accessibility Checklist." Version 1.1, March 1999. Available from `http://www-3.ibm.com/able/accessweb.html`.

Kao, John. *Jamming: The Art and Discipline of Business Creativity.* New York, NY: Harper Business, 1996

Microsoft Accessibility and Disabilities Web site. `http://www.microsoft.com/enable/`.

Nordic Cooperation on Disability. "Nordic Guidelines for Computer Accessibility," 1998.

Portuguese Accessibility Special Interest Group (PASIG). "PASIG Internet Accessibility Guidelines — Final Version," 1998. Available from `http://www.acessibilidade.net/doc/acessibilidade/guidelines_draft2.html`.

Internet Best Practices Standards Working Group. "Recommended Practice for Internet Practices — Web Page Engineering — Intranet/Extranet Applications." IEEE Std. 2001-1999, March 1999. Available from `http://computer.org/standard/Internet/webeng.htm`.

Blanchard, Harry E. *Standards: Standards for Multimedia, Accessibility, and the Information Infrastructure.* SIGCHI, Vol. 29, No. 3, July 1997.

World Wide Web Consortium. "W3C Authoring Tool Accessibility Guidelines," March 1999. Available from `http://www.w3.org/TR/WAI-AUTOOLS/`.

World Wide Web Consortium. "W3C User Agent Guidelines," March 1999. Available from `http://web1.w3.org/TR/WAI-USERAGENT/`.

World Wide Web Consortium. "W3C Web Content Accessibility Guidelines 1.0," March 1999. Available from `http://www.w3.org/TR/WAI-WEBCONTENT/`.

Chapter 4

Identifying Web Accessibility Barriers

In This Chapter

- Access technologies for people with disabilities
- Identifying the major barriers
- Problems for people who are blind or who have visual disabilities
- Web accessibility issues for people who are deaf or have hearing disabilities
- Facing the future of people with physical disabilities
- Additional disability communities affected by Web barriers

The publishing paradigm shift is nearly complete. We have moved from a paper-based, typewriter-generated, hand-edited, printing-press-produced publication toward a paperless, intelligent, WYSIWYG, software-generated, WWW-published hyper-document. Not only has the shift changed the way information is produced, but it has also changed the way individuals read that information.

For *temporarily able-bodied (TAB)* people, the shift has resulted in increased availability of a global information set. Unfortunately, because of this increased availability of information, the publishing industry has wrongly assumed that "one size fits all."

The sad truth is that the proliferation of information does not guarantee its accessibility. *Availability does not equal accessibility.* Where people with disabilities are concerned (particularly those with print disabilities), thoughtless barriers to information are being constructed by publishers of electronic information. The barrier factor is increased by the magnitude of inexperienced online businesses and organizations that have correctly assessed the inexpensive cost of delivering information on the Internet, but have inaccurately assumed that because it's on the Web, it must be easy to read . . . or access.

In this chapter, we consider the common accessibility barriers that people with disabilities must deal with each day. You will understand why the blind community is affected by Web accessibility issues the most today, and how people who are deaf, who cannot speak, or who have physical disabilities are likely to face greater challenges in the near future.

Let's first consider how people with disabilities access computers and electronic information today.

Access Technologies for People with Disabilities

Most people with disabilities require assistive or adaptive devices to help them render or view Web content. Those in the disability technology field refer to these devices or software interfaces by many names, including *access systems, assistive technology, adaptive technology,* and *adaptive computing.*

Examples of access technologies include the following:

- Synthetic voice, digital audio, or Braille for people who are blind
- Screen magnification and large text fonts for people with diminished vision or dyslexia
- Descriptive text, captioning, and visual cues for people who are deaf or people who have hearing disabilities
- Specialized adaptations for people who have physical disabilities involving the use of a keyboard, voice recognition mechanism, mouse, or other input device that requires a part of their body other than their hands and fingers to control a Web browser.

Specific types of Web access technologies for people who are blind or have visual disabilities include (but are not limited to) the following:

- **Screen magnifier** — Usually a software application that increases the size of text or images on a computer screen. Special monitors and other types of hardware adaptations can be used to project larger images as well.

- **Refreshable Braille display** — A hardware device that reads, translates, and subsequently renders electronic information from a computer interface to the user in Braille.

- **Screen readers and voice browsers** — Software applications combined with a synthetic voice that reads computer data back to users who are blind, or users who are more successful at auditory learning than reading due to a learning disability. This includes all screen objects (for example, windows and icons).

- **Synthetic speech** — Combined with a screen reader application or Web browser.

- **Speech recognition** — A software application combined with a speech input device (usually a separate or built-in microphone) that enables a blind user or a user with a physical disability to speak or issue commands that the speech recognition application recognizes and then acts upon. Speech recognition can also be used for creating and annotating existing material.

Web access technologies for people who are deaf or have hearing disabilities usually involve a captioning system or application combined with a player or plug-in (for example, RealPlayer) that can render the captions. If such a user is able to use his or her voice to speak, speech recognition may be a required access system. Generally, however, this is not the case. In Part II, you will see how several recent enhancements made to the Web help accommodate deaf users.

People who are physically disabled require a variety of access technologies, depending on the nature of their disability. For example, one of the fastest growing disabilities over the past ten years is repetitive strain injury (RSI). Interestingly, PC users make up part of the population segment often affected by RSI. The reason is obvious: Typing on a computer keyboard six to eight hours a day wrecks any person's hand muscles, nerves, and ligaments. (On a personal note, I've been pounding on a keyboard for nearly 20 years. Just recently I've noticed a significant amount of pain in my hands such that I now have to rest after 30–45 minutes of typing.)

Figure 4.1   Blazie Engineering's PowerBraille refreshable Braille display

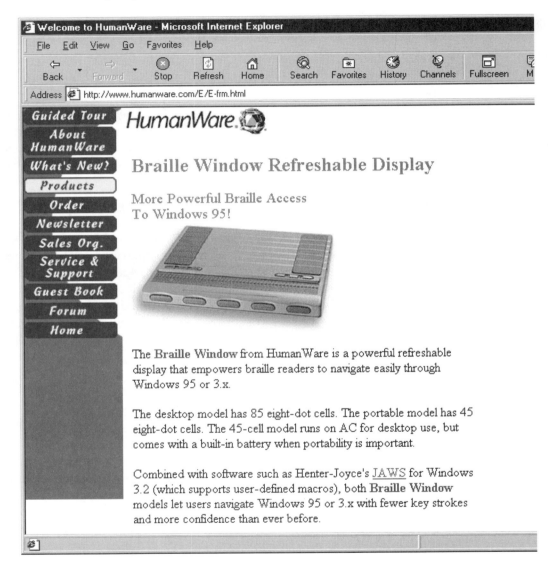

Accessible technologies for people with physical disabilities include speech recognition applications, head-pointing devices (see Figure 4.2), and eye trackers (also called *eye gazers)*.

Figure 4.2   The Prentke-Romicke HeadMaster Plus

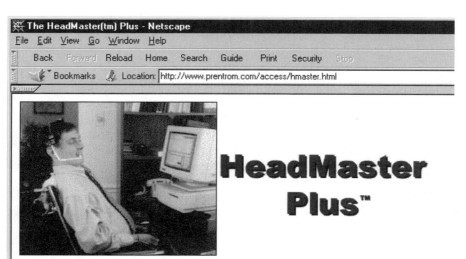

HeadMaster Plus is a head pointing system which provides full mouse control of com good head control. Moving one's head moves the cursor on the screen. Activating the p On-screen keyboards such as WiViK allow for word processing and other text entry.

HeadMaster Plus can emulate the most popular mouse input systems:

- Apple Desktop Bus for Macintosh™ and Apple IIGS™
- Microsoft™ mouse for IBM™ and compatibles with an RS232 serial port
- PS/2™ for IBM™ and compatibles equipped with the PS/2™ mouse port
- Sun Microsystems™/Unix

You can change the HeadMaster Plus from Apple to IBM compatibility just by flippin support for one-, two-, and three-button mice is built in. You can operate the mouse but switches into the back of the control box.

There are two points to note here:

1. Access system technology provides a way for people with disabilities to surf the Web and view Web content.

2. The accessibility of that content is crucial.

In theory, the designer of the information should not have to worry about producing several versions of specialized Web pages or sites. Rather, the focus should be on designing the source page with a rich set of characteristics that can subsequently be rendered or viewed by a wider audience — in this case, the communities of people with disabilities, or the previously mentioned TABs.

To emphasize, this not a new technology. Those involved in online publishing know that an electronic document can be coded using, for example, symbolic reference tags that are recognized by the document processor and then rendered to plain text, postscript, or browser-compatible output. The same preprocessing and postprocessing capabilities can be refined to produce Braille, large text, and synthetic-voice output documents. No doubt with the advent of publishing mechanisms such as CSS2, SMIL, XML, and MathML, we are reaching the point where natural language voice documents (NLVDs) are possible.

As a result, access is not only achieved for people with disabilities, but language barriers also diminish. Indeed, with the W3C and commercial support for style sheets, the ability to produce accessible information for all people — disabled or not — has never been greater.

However, as you will see in the next few sections, in spite of recent advancements, barriers still exist for many people with disabilities.

Identifying the Major Barriers for Those with Visual Disabilities

Several barriers and difficulties involving access to the Web exist for those with visual disabilities (the term *visual disabilities* includes anyone who has a sight difficulty, including people who are blind).

In the information design phase, the most common issues related to Web site inaccessibility involve the following areas:

- Complex notation
- Image rendering
- Multimedia
- Frames

- Forms
- Tables
- Navigation
- Java scripts and programming code

The blind community is affected by each of these areas, sometimes severely.

Science and Math Notation

Complex notation, including math and science notation, is extremely difficult to render in an acceptable format, especially for people who are blind. Note that most blind people require an alternative format of the information. Likely the alternative is ASCII text, Braille, or synthetic speech. Even these alternatives do not guarantee absolute accessibility. For example, not all languages contain characters that are represented in ASCII. Therefore, the proper rendition — that is, the proper functional order of the notation — is critical to the reader. Recent advances at the World Wide Web consortium involving the Mathematical Markup Language (MathML), however, do give cause for hope. MathML uses Unicode, which contains a much larger character set to identify each character whose visual appearance (glyph) is intended.

However, the challenge of rendering complex notation that tends to be graphical in appearance often requires additional hand editing in order to complete the transformation process. Few Web publishers are interested in making that investment. Time is a commodity, and being forced to take extra time to ensure that notation is accessible is not something Web content producers want to invest in.

Consequently, Web content such as science, math, and business educational material is often inaccessible to a blind person. When you consider how many post–secondary schools are moving to Web-based educational systems or require research where the Web is the research tool, you can appreciate what a blind or visually disabled person has to deal with.

One of my colleagues and the technical editor for this book, Harvey Bingham, provided me with the following list of challenges people with visual disabilities face trying to read math:

1. The American Math Society has included 2,322 named glyphs in MathML. The meaning given to these glyphs is part of the semantics of the mathematical writing, which is up to the author. The glyph names

may suggest the geometrical appearance (for example, "uprightharpoon"), but not the author's meaning.

2. A linear textual description of an equation using MathML mathematical notation, using the glyph names, is much longer than the visually presented equation. It needs semantic grouping aids to chunk the description for study and understanding. Determining the relationships and navigation among the chunks is important.

3. A linearized presentation of the math removes the two-dimensional visual image that helps to show the author's intended structural representation. Many linearizations of a visual image are equivalent, and there is no canonic representation. So the visualization from the linearization needs to keep the author's structural intent and assigned semantics.

HP EzMath, an application created by Dave Raggett and Davy Battsalle, is a great example of how to code mathematical expressions so they render properly. You can find HP EzMath at `http://www.w3.org/People/Raggett/EzMath/`.

Images and Image Maps

Image rendering presents similar problems. Because the blind or low vision user is likely to be using alternative output (for example, a screen reader), an image that does not include a textual description becomes a barrier. Additionally, the textual description needs to have contextual meaning.

To illustrate, if your Web page contains an image of your corporate logo and you include the text description "image," what good is that to the user? It doesn't convey useful meaning at all. When your graphic design team created the logo, they no doubt invested a considerable amount of time to ensure that the logo conveyed meaning and corporate identity. Do you think that a Web surfer wants less? Hardly. Now put yourself in the place of a person who visits your site and is greeted by an image map that functions as a navigation bar throughout the site. If you code that page so that each image link does not contain an alternate text description, a person with a visual disability will find it next to impossible to navigate your site. Using synthetic voice output, the blind user would only hear the word "image" repeated over and over again. This is not an unusual experience for blind Web surfers. Remember too, it's not just the blind user you are catering to. Audio browsers, self-voicing browsers, phone browsers, and text browsers are becoming increasingly popular. They require textual and/or aural content, not graphical content.

I

4

Therefore, design considerations must be implemented. For example, using meaningful descriptive text in conjunction with figures, images, or other graphical entities within a Web site. Descriptive or alternative text must become a standard attribute for Web-based documents.

To encourage the use of alternative text, some authoring tools like Dreamweaver (http://www.macromedia.com/) and Corel WEB.SiteBuilder have built-in mechanisms for alternative text prompting. However, these same authoring tools are generally not accessible to the blind user. They have not taken advantage of the programming hooks that enable access technologies such as screen readers to render the application accessibly to the blind. In other words, it's bad enough that content isn't accessible, but the publishing tools themselves are not accessible.

The first implementation of graphical browser support for alternate text is Microsoft's Internet Explorer. IE automatically displays alternative text using bubble help–type notes as the user passes over the image with the mouse cursor. Not long after this implementation, Netscape followed suit and Navigator/Communicator now also displays the alternative text of an image or image map. When IE or Netscape Navigator are used in conjunction with windows screen readers for the blind, the alternative text is easily rendered. A sighted user can also use this mechanism to quickly scan a page to determine whether the ALT text is available.

Descriptive Video

Descriptive video provides a blind or low vision user with additional narrative that is useful and sometimes critical to their comprehension of an electronic document. The process simply requires the interjection of descriptive narration during the spots within the video that are not otherwise filled with sound effects or dialogue. As a result, the blind or visually disabled viewer achieves increased comprehension of the video event.

The National Center for Accessible Media (NCAM) in Boston, Massachusetts currently provides a service that implements descriptive video for the motion picture industry. Descriptive video and captioning (for people who are deaf) are perfect examples of how the power of markup should be used to enhance the richness and accessibility of a Web page. They have recently received grants to assist them in the research, design, and delivery of Web-based information for public television.

New aids in the advancement of descriptive video and captioning include the W3C's Synchronized Multimedia Integration Language (SMIL) and Microsoft's Synchronized Accessible Media Interchange (SAMI) tool.

Navigation of Forms, Tables, and Frames

Navigating a Web site, particularly a hypertext document, is a challenge for anyone. Keeping track of where you've been, where you want to go, and then getting there can be a cyber-nightmare. Still, being able to visually navigate through a Web site has obvious advantages that the blind or low vision user cannot easily imitate. A navigational cue as simple as colored text provides meaning and definition that the nonvisual user cannot see. Therefore, there is a need to design solutions that implement audio cues in concert with visual cues.

Remember, too, that navigation is often closely tied with memory and consistent design. People with cognitive limitations simply require visual memory aids and simplified page design. An example of this can be found at the WebABLE! Web site (http://www.webable.com/). The designer, Colin Moock, implemented a system of visual cues consisting of opened and closed doors. The concept is basic to most people and simple to learn and remember. Underneath each image is the associated alternative text for a blind person to easily follow.

Navigational difficulties clearly present challenges to every user. Consider the difficulties a visual user experiences with the Web today. Then imagine trying to surf the Web with your computer monitor turned off — the only way to navigate being to listen to a description of a Web page. The challenge is incredible.

Tables

Another classic barrier is the design of Web pages with tables. In Chapter 5, I cover how important it is to separate structure from presentation. Suffice it to say this is a critical design phase because many blind Web surfers use screen readers to read Web pages.

If a Web site is constructed using tables (and thousands are), the blind user can easily become confused. Why? Because screen readers are only capable of reading lines of information from left to right. They cannot distinguish or interpret a Web page designed with tables and loaded with different information sets in different table cells or embedded tables with multiple table columns and rows. From the visual user's perspective (as well as the graphic designer's), this may appear to be the only way to place objects or text on a Web page exactly where he or she wants them. However, this is neither the proper use of the HTML table syntax nor the only way to specify locations on a Web page. Style sheets are a better method.

Forms

A similar problem occurs when a Web author designs a complex online form that contains multiple groups of form input boxes but does not contain accessible HTML syntax to simplify the process of filling out the form. A classic example of an inaccessible type of form is a Web subscription form for an online journal. Navigating this kind of form is extremely difficult for a person who is blind, especially if they make a mistake and have to go back to correct something. This is due to a lack of shortcut keys, form controls, and grouping mechanisms, all of which are part of the HTML syntax. The Web author can aid the user by providing descriptive text in input areas to suggest what is needed there. The user then need only modify or delete what is there.

Frames

Frames aren't just a navigational challenge for the blind Web surfer; they tend to present a total comprehension problem. The use of frames may make it easier for the Web designer to present multiple windows of static information, but it cuts against the grain of the standard human reading paradigm. I can think of only two successful uses of a frame-type technology: picture-in-picture televisions and online help systems. Even in these examples, a person still requires a focal point. Web frames traditionally throw that logic out the window.

Blind people using screen readers or self-voicing browsers can identify the number of frames presented on a Web page. And, if properly coded (using the *title* attribute), they may be able to tell you the general contents of a frame. But their user agents (browsers and screen readers) cannot follow multiple events in multiple frames. They can only follow one frame at a time — the frame they currently reside in. Therefore, if your Web site displays important information (even if it's repetitive) in different frames, chances are the blind Web user cannot capture it.

Without a doubt, navigation requires acute sensory awareness. Navigation is not just a document road map; it is not a linear link. Rather, good navigational design includes a combination of seeing, hearing, and feeling your way to a specific destination in a comprehensive way.

Web Accessibility Issues for People Who Are Deaf or Hard of Hearing

Web sites that contain multimedia features, including sound and video clips, require additional attention. Keep in mind that anything emitting sound cannot be heard by the deaf. It also may not be heard by the hard of hearing or, for that matter, anyone viewing the site in a noisy environment. If you believe this is an impractical example, consider the current industry move to WebTV and Web kiosks in public facilities.

However, opportunity always implies challenge. One of those challenges is found in the current evolution of the Web — moving from a text-based interface to a multimodal, multimedia operating environment. It is this environment that presents barriers to the deaf.

For example, each time a Web site includes a video clip that also includes sound without text description or captioning, people who can't hear are locked out.

The solution is relatively simple: implement closed captioning. You may know that captioning is now an industry standard for televisions. As a result, you see televisions with built-in captioning functionality. Deaf users are no longer required to purchase a separate captioning box. For general information about captioning, please refer to the Closed Captioning Web site (http://www.captions.org/).

Previous implementations of Web captioning were not the same as television captioning. Rather than captioning within the video clip, Webmasters included captioning or script indicators on their pages. These indicators were located in close proximity to the video clip hyperlink (or image). When a user clicked the captioning indicator, the script of the clip was displayed.

Recent advancements in Web captioning have been developed. Applications like QuickTime (http://www.apple-imac.com/quicktime/) and SAMI (http://www.microsoft.com/enable/sami/details.htm) are both capable of embedding captioning in video and audio Web segments.

One of the best sites discussing captioning is that of Boston-based public broadcasting station WBGH (http://www.wgbh.org/wgbh/pages/ncam/). Among many beneficial services provided for people with disabilities, WGBH runs the National Center for Accessible Media (NCAM), directed by Larry Goldberg. NCAM provides complete instructions for implementing Web-based captioning (http://www.wgbh.org/wgbh/pages/ncam/webaccess/captionedmovies.html).

Following are some guidelines you should consider before implementing captioning on your Web page:

- Be sure to inform users that you have implemented captioning. Providing a specific set of instructions on your home page is ideal.

- Try to use a textual captioning indicator rather than an image or icon. This makes it easier for the visually or hearing disabled user to access the indicator. If you do use an image, be sure to provide alternate text.

- Remember to include the size of the video file. This is a usability measure that assists all users. Likely you will find that many users preview the script before they download the video clip.

Increased accessibility on the Web for people who are deaf or hard of hearing can also be improved by ensuring that all emitted messages (error or information, system or application) are displayed through visual cues as well as audible cues. This is particularly important for browsers, authoring tools, and public kiosks.

Facing the Future of People with Physical Disabilities

For people with mobility disabilities, accessibility issues can take on a wide range of challenges. Some people have use of their hands, while others do not. Some have the ability to use mouth sticks and head pointers, while others rely on infrared devices. Still others appear to have no barriers present when their interaction with the Web is through a personal computer. However, when faced with a public Web kiosk, these same users may be presented with inaccessible physical control options.

Additionally, with the recent integration of voice recognition on the PC platform, I suspect that it will not be long before Web interfaces implement a similar user interface. If and when this occurs, it will certainly present a series of problems for the people with physical disabilities that have difficulty speaking. Certainly this includes those who are not able to speak and those who are deaf.

While the Web may not present as many major barriers to people with physical disabilities, presentation of content should be given some credence. Because of various physical difficulties, head and eye movement are not always easily accomplished. Keep the following guidelines in mind:

- Maintain a simple design that is easy to view.

- Be sure to provide onscreen navigational controls that are easily identified. People who require the use of an assistive device such as a head pointer (a stick-like device attached to the user's head to enable keyboard typing) need to be able to easily access those controls.

- Browser developers should provide keyboard mappings to all browser operations.

- If you are designing a Web application that runs on a public kiosk (for example, in libraries, museums, or government agencies), the kiosk itself should be accessible to a person using a wheelchair. Kiosk height, control knobs or buttons, and input mechanisms should be easily accessible.

Additional Web Accessibility Barriers

Web content presents a major set of barriers to people with disabilities. But you should know about some additional barriers involving Web clients (user agents and browsers) and publishing tools.

Publishing Tools

The accessibility issues related to the production of a document primarily involve the following:

1. Authoring tools that are themselves inaccessible and fail to enforce or prompt for accessible design tags, semantics, and programming syntax
2. Publishing tool processors that are not capable of generating accessible documents from the original source file
3. Publishing tools that remove attributes needed for accessibility

SoftQuad Software's HoTMetaL PRO is the first Web site development tool that can build accessible Web pages because it provides both accessible HTML prompting and includes accessibility HTML validation. Recently, Chami.Com (HTML-Kit), Sausage Software (HotDog Pro), Allaire (Home-Site), Adobe (GoLive), and Microsoft (Frontpage) have introduced measures of accessibility into their products. This is a great sign of things to come and an indication that industry is taking the issue of accessibility seriously. Note that the W3C has also improved the accessibility of Amaya, their web browser/authoring tool.

Of course, anyone can produce accessible HTML files if they write them using a text editor such as Windows Notepad rather than a tool like HoTMetaL PRO that is specifically intended for the creation of HTML files. Other tools may also enable you to edit the code they produce, and then leave it alone when you're done — in other words, the tool won't go back and edit it for you when you load it on another occasion.

However, SoftQuad Software built accessibility support into HoTMetaL Pro gives from the get-go. It is usable by blind people using screen readers as well as people with mobility disabilities who require HoTMetal's visual dynamic keyboard (VDK). I find this a laudable accomplishment by Soft-Quad, and hope that other tool publishers quickly follow suit. No other tool provides this level of accessibility support

---

Cross-Reference

See Chapter 8 for more information about publishing tools and accessibility.

Browser Barriers

Web pages are gradually becoming more accessible and requiring less massaging by a postprocessor or other intermediary actions. This is because some Web browsers contain accessibility features that enhance the readability of a document to people with disabilities. However, the features are minimal at best and often require the addition of an assistive technology to enhance the accessibility of the page. For example, Internet Explorer and Netscape Navigator are much more accessible when used with a screen reader for the blind.

Alternative browsers are available. Good examples include pwWebSpeak from The Productivity Works, Inc. (http://www.prodworks.com/) and HomePage Reader from IBM (http://www-3.ibm.com/able/hpr.html). These browsers were designed with synthetic speech and large text functionality built in. They also support the W3C's HTML 4.0 specification that includes several accessibility enhancements.

Accessible Web page delivery through a browser can be significantly enhanced if developers include assistive preference options that enable a user to turn on captioning, descriptive video, sound cues, synthetic voice, keyboard mapping, screen magnification, and other accessibility features.

Summary

In this chapter, you learned the following:

- Web surfers who are blind or have visual disabilities are affected by inaccessible Web content, clients, and publishing tools.
- Users who are deaf or hearing impaired are affected by the lack of captioning and Web description of multimedia.

- People with physical disabilities may be presented with barriers involving Web devices, including WebTV and Web kiosks.
- Content producers are not using available tools or HTML 4.0 accessibility syntax to increase the accessibility of their Web pages.
- Vendors of browsers have failed to include preference settings that could significantly improve access for people with disabilities.
- Most publishing tools are not only inaccessible, but also do not include accessibility prompting or validation and may remove already existing accessibility attributes or prevent them from being added.

The next chapter examines HTML coding techniques you can implement to create accessible Web pages. You will see how, with minimal effort, you can create a Web site that people with disabilities can easily surf, thus opening the doors to universal accessibility.

References

Battsalle, Davy and Dave Raggett. "Add Math to Web pages with HP EzMath v1.1," 25 October 1998. Available from `http://www.w3.org/People/Raggett/EzMath/`.

Bachrach, Steven M., Peter Murray-Rust, Henry S. Rzepa, Benjamin J. Whitacker. "The Chemical Markup Language." (From the article "Publishing Chemistry on the Internet.") Available from `http://www.netsci.org/Science/Special/feature07.html#cml`.

World Wide Web Consortium. "W3C MathML Specification, Version 1.0." Available from `http://www.w3.org/TR/REC-MathML/`.

Part II

Accessible Web Site Design

Chapter 5

Creating Accessible Web Sites with HTML

In This Chapter

- Introduction to accessibility design issues
- If you already have a Web site
- If you are creating a new Web site
- Summary of HTML 4.0 enhancements for accessibility
- The importance of separating structure from presentation
- Describe everything
- Increase efficiency
- Creating accessible forms
- Frames and tables

Many Web designers subscribe to the notion that providing for accessibility in their Web sites is too expensive, too time-consuming, or just plain too difficult. This book, the Web Accessibility Initiative, and other programs are working diligently to prove those assumptions wrong. This chapter introduces you to techniques that can simplify your approach to making your site as accessible as possible.

Introduction to Accessibility Design Issues

It has often been said that accessibility represents "the curb-cuts on the Information Superhighway." Though some might feel the Information Superhighway terminology is tired, the metaphor most definitely rings true.

What does the average person think about when faced with the term *accessibility?* Disabled parking stalls, roomier restroom accommodations, wheelchair ramps, Braille instructions on elevators, and of course, curb-cuts in the sidewalk.

Do these features of public life only improve the quality of public access for the disabled? Or have the rest of us found them to be quite an enhancement as well? You may be old enough to remember the days of lugging a child and stroller up over the curb at each block, or up the stairs at city hall — all while carrying the ever-present diaper bag and any packages or papers needed on your errands. Delivery personnel struggled dozens of times a day with heavy loads on hand trucks that had to be hefted those four to five inches per stair at each delivery. Trips to public restrooms with preschoolers could be an exercise in flexibility, as you tried to fit both of you into the stall and undress the child without resorting to standing on the plumbing fixtures.

Sidewalk curb-cuts, ramps that bypass stairs, and larger restroom facilities help many of us in life's daily tasks, not just those who use wheelchairs and walkers. Is it possible that the same result could apply to electronic curb-cuts?

What Is Accessible Design?

Accessible design begins with an accessible designer. Many Web page designers innocently envision other users being much like themselves. If you aren't exposed to people with disability issues on a regular basis, it's easy to forget that not everyone has the same physical capabilities. The accessible designer remembers that among other things, Web accessibility is designed to promote access by individuals with varied environments. Following are some things to keep in mind about potential users from the W3C Web Content Accessibility Guidelines:

• They may not be able to see, hear, move, or understand easily — or at all.

- They may be in a situation where their eyes, ears, or hands are busy or interfered with (for example, driving to work, working in a loud environment, and so on).
- They may have difficulty reading or comprehending text.
- They may not have or be able to use a keyboard or mouse.
- They may have an early version of a browser, a different browser entirely, a voice browser, or a different operating system.
- They may have a slow connection, a small screen, a text-only screen, and so on.
- They may not speak or understand fluently the language in which the document is written.

Each of these situations lends itself to the original idea that Tim Berners-Lee had for the Web when he wrote his original proposal at CERN: Any user, anywhere, at any type of terminal, should be able to access information. By creating Web sites according to today's accessibility guidelines, authors continue to fulfill that ideal.

Guiding Principles

Accessible design practices are broken down into a concise set of 14 checkpoints in the Web Content Accessibility Guidelines.

Note

Web accessibility conformance is predicated on browser recognition and implementation of strict HTML 4.0 (http://www.w3.org/TR/REC-html40/). Not all browsers have reached that level and, subsequently, some browsers do not support all of the HTML 4.0 accessibility enhancements.

Provide equivalent alternatives to auditory and visual content

All images and area attributes within image maps should have alternative text provided in the ALT attribute. Not only does this make the Web page accessible, but it is also required by the HTML 4.0 Recommendation. Image-based navigation tools and form elements all need text-based alternatives such as a full text-navigation menu.

Important visual information can include photographs of a retailer's products, a graph or chart of financial information, or a video clip of a news story. Each of these objects should have an additional text-based alternative presentation of its content. This long description differs from the alternative text in the previous guideline, as it is a reformatting of the visual content, not a description of it. For example, an alternative text for a video clip might appear as follows: ALT= "news video story on recent airline disaster." However, the LONGDESC content would be a URL pointing to a file containing a transcript of the audio portion of that news feed, along with a description of any key visuals. You can also use LONGDESC to implement a text transcript for motion pictures or videos that include Descriptive Video Services (DVS), which are full audio descriptions of spoken dialog and sound effects.

Note

Implementation of the LONGDESC attribute unfortunately doesn't exist in the 4.0 versions of today's most popular browsers.

Transcripts should always be provided for audio-based content. This includes audio portions of video, as well as descriptions of important but nonverbal audio communication. If a sound file of the male lion's roar accompanies the presentation of a photograph of a pride of lions, describe that to the user. Not only does this serve users with hearing disabilities, but it can also help users who are at workstations without sound cards, or who may be struggling with a slow Internet connection. Additionally, search engines can pick up this text-based content.

Don't rely on color alone

Color should never be the only indication of importance or context for your information. Users who are color blind or are using a device that isn't capable of color display are not able to perceive such emphasis.

Use markup and style sheets properly

Avoid using structural markup for the visual result given by a few popular browsers. Documents that contain orderly and appropriate structural markup can transfer to almost any type of user agent or display device and still stand on their own.

Clarify natural language usage

Use the new markup for abbreviations and acronyms to help expand text that may be unfamiliar to your reader. Also, identify changes in the natural language of a document where they occur by using the LANG attribute.

Create tables that transform gracefully

Tables should be reserved for truly tabular data; they should not be used for fine control of visual layout (such as placing text into columns). One metric is to read the information in your table horizontally, without regard to column borders. If it makes sense, it's tabular data. Otherwise, you've used the table for layout rather than structure.

II 5

Ensure that pages featuring new technologies transform gracefully

Being on the cutting edge can be fun, but be aware that many of your users may not have a user agent capable of rendering new content formats. Be sure to provide alternative formatting for content contained in frames and scripts by using the NOFRAMES and NOSCRIPT elements. Any dynamic content should have an alternative presentation that does not rely on the interactive capabilities of the user agent.

Ensure user control of time-sensitive content changes

Not everyone can read or process information at the same pace. If you have scrolling or changing information, be sure there is a mechanism available for the user to pause or stop the flow of data.

Ensure direct accessibility of embedded user interfaces

If an embedded object such as an applet or ActiveX control is used, the interface for that object must also be accessible. Details on such techniques are discussed in Chapter 9.

Design for device independence

Not everyone has a mouse, and not everyone has a keyboard. Keyboard shortcuts, tabbing order, and event handlers all contribute to the accessibility of forms and elements. Keep in mind that people with a variety of input devices may interact with your documents.

Use interim solutions

Accessibility may simply mean being thorough in your implementation of design elements, making allowances for browsers that don't fully or correctly implement elements. For example, by using placeholder text in form text-input controls, an older browser enables users to navigate directly to that control, where they have difficulty doing so by tabbing, for example, with empty elements.

Use W3C technologies and guidelines

Adhere to W3C recommendations for markup and content development guidelines; this provides greater security that your documents degrade gracefully to software that may not fully implement the most recent technologies.

Provide context and orientation information

Complex pages can be daunting to even the quickest user. Group related elements together using labels, headings, and names for frames. Form elements should be clearly labeled and placed into option groups.

Provide clear navigation mechanisms

The most frequently used feature of any Web site is the navigation system. No matter where on a site a user may be, there should be a way to return to the home page or major subsections with only one or two steps.

Ensure that documents are clear and simple

Consistent layout and navigational icons aid those with cognitive disabilities. Clear and concise language can assist all users, and can be a great help for those whose native language is not that used on your site.

If You Already Have a Web Site . . .

The prep-work involved in retrofitting an existing Web site consumes more time than any other task. However, completing this preparation properly is the biggest key to the success of your project.

Take Inventory of Your Site's Content

It sounds simple, but before you can edit your site, you must be sure you know what's there. The type of content present may impact the decisions you make about navigation systems, organization of data, and the presentation of major elements.

Make a list of each occurrence and each file name for the following items:

- Images (informational or decorative)
- Forms
- Java applets, ActiveX controls, or other programmatic objects
- Inline scripts (JavaScript, VBScript, and so on)
- Framed content
- Dynamically generated content

Images

Each image should be evaluated for purpose. Does it serve purely as decoration? Is it part of a visual navigation system? Does it convey information that isn't addressed in the document text?

After categorizing your images, determine which ones can be retained and which ones could be more appropriately converted to text. Those that remain require an ALT attribute for alternative text. Images that convey information not found elsewhere in the document should have a long description developed and linked with the LONGDESC attribute.

Forms

Analyze each form to ensure controls are grouped logically. Each control must have an associated label element in addition to any textual cues as to the intended content.

Java applets and other programmatic objects

Java applets and other programmatic elements are tempting to use for dynamic or interactive functions. It's important to remember, however, that these elements have a user interface of their own that requires as much attention to accessibility as your Web page does. (Chapter 9 addresses programmatic accessibility in detail.) In terms of content accessibility, the information presented or solicited must be available through alternative means.

Inline scripts

Scripts that provide content rather than visual effects — such as mouseovers — require an alternative means of information delivery for user agents that don't support scripting languages. For example, a JavaScript-based "calculator" (often found on financial sites) could have a link to a description of the formulas used in the computations, or to a table of possible results.

Framed content

Every frameset must have an accompanying NOFRAMES element for user agents not capable of or not set to render framed content. Additionally, the content provided through NOFRAMES must still make sense when viewed outside of the framed environment.

Dynamically generated content

Do the pages require specific user intervention before rendering, or is the process transparent to the user?

Check Each Inventory Item for Compliance

Download or print the list of checkpoints for the Web Content Accessibility Guidelines from the WAI program Web site at http://www.w3.org/TR/WD-WAI-PAGEAUTH/full-checklist.html. Work through each segment of your site inventory and apply the checkpoints pertinent to that element. For example, with each image, write text for the ALT attribute and provide a link to a long description using LONGDESC.

Put the pieces back together and review the entire site from a big picture perspective. Having started small with the individual element, you're now in a good position to review the site's global issues. Does it continue to pass the items in the checklist? Do any additional changes need to be made to enhance navigation, or provide alternative access to major portions of the site?

If You Are Creating a New Web Site . . .

A new Web site gives you the perfect opportunity to provide a rich and rewarding experience to all of your visitors: able-bodied or with disabilities, those using traditional visual browsers, or users with screen readers, hand-held PCs, or even cellular phones.

Refer frequently to the guiding principles of accessible design listed in this chapter, which are taken from the WAI W3C Web Content Accessibility Guidelines. Consider printing the list of checkpoints for the W3C Web Content Accessibility Guidelines published by the WAI program, which is available at http://www.w3.org/TR/WD-WAI-PAGEAUTH/full-checklist.html.

Plan Effectively

You now know that having a clear and effective navigation system is a cornerstone of good accessible design. When planning the flow of information across your site, always think of how each page or section relates to the others, and how a user navigates between them.

Summary of HTML 4.0 Enhancements for Accessibility

The W3C's recommendation for HTML 4.0 was the first time authorities in Web accessibility were invited to participate in the process of producing and reviewing the working drafts. The W3C's own Web Accessibility Initiative coordinated this effort, bringing together work from formal working groups such as the Page Authoring Guidelines Working Group, and informal sources such as mailing lists for those interested in accessibility issues.

The WAI program sought remedies to four major problems:

1. Authoring habits that cause problems for users with screen readers, audio browsers, and text-only browsers

2. Unstructured pages that result in confused users and barriers to navigation

3. Techniques that rely upon structural elements for formatting or presentation, adding a semantic to the markup that aren't implied by the HTML recommendations

4. Overuse of graphical information without text-based alternatives

To these ends, the improvements made in HTML 4.0 are highlighted in the following sections.

Additional Structural Elements

The HTML 4.0 Recommendation brought with it many new elements that gave Web authors additional structural components to work with when blocking out documents. These include the following:

- For structured text, the ABBR and ACRONYM elements, in conjunction with style sheets and the LANG attribute for indicate a change in a document's natural language. For individuals using speech-based browsers, these constructs make it easier to identify the meaning of the abbreviation or acronym, as well as identifying language changes in text.

- The `BLOCKQUOTE` method of segregating quoted material has been enhanced with the new Q element for inline quoting.
- `INS` and `DEL` are new elements to mark insertions and deletions to text. These features are important for continuity in legal documents and other areas where keeping track of the trail of changes is imperative.
- Improvements in forms include logical grouping of elements and labels, all of which are discussed later in this chapter.
- Tables have vast improvements in structure, enabling headers and footers to be set aside as specific structures so that user agents can render them in meaningful ways. Additional attributes include scoping and axes for better correlation of data by nonvisual user agents.

The Importance of Separating Structure from Presentation

When planning a new Web page, many authors first think about what it will look like. While visual presentation is important, to the accessible designer that's putting the cart before the horse, so to speak.

The structural needs of the document determine which HTML elements are needed. Review the printed version of the document — or if the document doesn't exist yet, create a list of major components to be included in the document. Label each component by its function — for example, "this is a heading," or "passage quoted from another work." Identify which HTML elements are appropriate for marking up each component.

When you've finished, review your labels and proposed HTML elements to determine if they provide structure or presentation. A header, such as this section's "The Importance of Separating Structure from Presentation," provides structural division within the document content. That the heading may be rendered in bold, 18-point Arial font is the presentation. Reannotate your document as necessary to separate the structural intent of a section from the presentational markup.

 Tip

The WAI Techniques for Web Content Accessibility Guidelines document has a convenient matrix of elements that are considered structure vs. presentation. You can find it at `http://www.w3.org/TR/WD-WAI-PAGEAUTH/wai-pageauth-tech.html#structure`.

Once this process is completed, you'll have a basic structural framework for your document much like an outline for a report or the floor plan for a new house. You can then take that structure and present it in whatever fashion works best for the user. To extend the metaphor of a new house, the presentation of that structure in wildfire-prone areas of California likely includes a stucco exterior finish and Spanish tiles on the roof. In contrast, a resident of Florida is most concerned with the extra support that concrete block gives during hurricane-strength winds. By providing a sound structure to the plans, the developer can easily accommodate both owners' needs without changing the basic integrity of the design. So, too, can the Web developer!

Describe Everything

Any component of your Web site that provides visual information cannot be understood by those who cannot view it. Not only does this mean the visually impaired, but also any user who may be accessing your site with a browser or device that cannot fully display images, objects, or form elements. To ensure that these users are presented with as much information as any other user, HTML provides for descriptions that may be passed to the browser in place of the troublesome element.

The ALT Attribute

The ALT attribute is intended to provide an alternative text equivalent to the graphical information being presented. A trend in some development circles has been to place the file name and image size in the ALT attribute. While some users may find this helpful, it does not qualify as a text equivalent to the information being presented. Instead, use text that literally represents the image. The Trans World Airlines Web site logo in Figure 5.1 illustrates this technique quite clearly. The logo reads "Welcome to TWA.com." The ALT text? "Welcome to TWA.com." Simple, isn't it? The HTML markup for this could look like the following:

```
<IMG SRC="welcome.gif" ALT="Welcome to TWA.com">
```

Figure 5.1   ALT **text as a direct equivalent to image content**

ALT attributes aren't only applicable to individual images. They should also be applied to image maps and applets. Any graphics-based conveyance needs equivalent text for all users to have access to the same information.

The LONGDESC **Attribute**

Images, image maps, and applets all present significant information visually. Consider how you might create a Web page discussing a new painting you just hung in your home. The description might include the following:

```
<BODY>
<P>I just found a fabulous new painting. Here's a picture of it hanging
on the wall in my living room. <IMG SRC="photo.jpg">
</BODY>
```

When viewed by someone using the nongraphical Lynx browser, this page would be rendered as seen in Figure 5.2.

Figure 5.2 **Images without** ALT **text provide no useful information to nonvisual user agents**

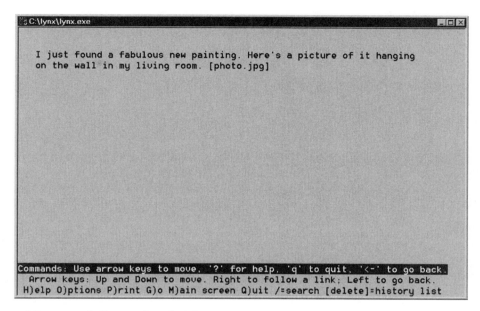

Not very informative, is it?

The ALT attribute can be used with the IMG element to provide text that may be displayed instead of the image when a user agent isn't capable of rendering the actual graphic, or a user has chosen to surf with images turned off in a visual browser. The alternative text presented should be chosen carefully.

```
<IMG SRC="photo.jpg" ALT="My new masterpiece">
```

This IMG element does indeed include alternative text, but it's almost as meaningless as not having any at all. The context provided with the original version already told the user that the image was of a new painting. The alternative text describing it as a "new masterpiece" does nothing to enhance the user's experience of the page. Consider using the new HTML 4.0 attribute LONGDESC and coding your Web page as follows :

```
<IMG SRC="photo.jpg" ALT="photograph of a painting" LONG-
DESC="photo.html">
```

where `photo.html` contains the text:

```
The image is a photo of 'Waterlilies' by Monet. Delicate, floating
flowers produced in oil-based paint on canvas in a gilded frame.
```

This version provides a concise description of the content of the image, enabling all users to appreciate the new addition to your art collection.

As previously noted, not all browsers support strict HTML 4.0. The `LONGDESC` attribute is one of the key constructs not supported by any commercial browser. However, there is a way of coding your HTML to assimilate the function of `LONGDESC` which commercial browsers do support. This is accomplished by creating a description link near the graphic. Continuing with the example above, here is how you would code your Web page to create the descriptive link:

```
<IMG SRC="photo.jpg" ALT="photograph of a painting" LONG-
DESC="photo.html">
<A href="photo.html" title="Description of Painting">[D]</A>
```

Subsequently, when the Web page is rendered in your browser an uppercase letter *D* is displayed near the graphic functioning as a hyperlink to the file `photo.html`. The default color of the hyperlink is blue, but it may be different depending on the settings of your Web browser or the color properties established by the Web page author.

This technique was first developed by the staff at the CPB/WGBH National Center for Accessible Media (NCAM) (`http://www.wgbh.org/wgbh/pages/ncam/`). It is a feature available in SoftQuad Software's HoTMetaL PRO Web publishing application (`http://www.sq.com`). Refer to Chapter 8 for more information about HoTMetaL PRO and how to implement the description text link.

The `TITLE` Attribute

Web authors are familiar with the `TITLE` element, which provides a short descriptive name for an entire document. The `TITLE` attribute, on the other hand, provides additional means for describing elements such as frames and images. Visual user agents often interpret the `TITLE` attribute as a tool tip — a bit of text that appears when the pointing device hovers over an item on screen (see Figure 5.3).

Figure 5.3   A title displayed as a tool tip by a visual user agent

While TITLE and ALT may seem redundant for images, user agents may interpret them in different ways. Besides images, the TITLE attribute provides for the expansion of abbreviations and acronyms when used in two new elements, which I discuss next.

The ABBR Element

A common problem with informative text is defining new terms or abbreviations. Most publishers provide a visual offset, such as italic font, at first use. HTML provides the ABBR element to do the same for abbreviations. The syntax for the abbreviation element makes use of the TITLE attribute discussed previously. This reinforces the concept of describing everything for your users. For example, it is very helpful to a blind user who is reading a Web page using a speech-based browser because it clarifies the context of what is written, particularly for unfamiliar abbreviations.

To markup the abbreviation "etc." use the following structure:

```
<ABBR TITLE="building">bldg.</ABBR>
```

The ACRONYM Element

The ACRONYM element is functionally similar to the ABBR element, as it provides a definition for the acronym contained within. The important distinction is using each element appropriately. For example, HTML is an acronym for "Hypertext Markup Language." Bldg., on the other hand, is an abbreviation for "building."

Similar to the concept of properly interpreting text marked up with the ABBR element, a speech-based browser supporting the strict HTML 4.0 specification is able render the meaning of an acronym to an individual who might not otherwise recognize the term when spoken.

```
<ACRONYM TITLE="HyperText Markup Language">HTML</ACRONYM>
```

The LANG Attribute

Information about the language being used is also often presented visually in printed text. If a mode of dress were to be described as de rigueur, most publishers would italicize the phrase, as seen here. To provide meaning in addition to the visual offset, the LANG attribute was developed. It may be applied to any element that contains text.

```
<P>The United States quarter dollar coin is imprinted with the phrase
<SPAN LANG="LA">E Pluribus Unum</SPAN>.
```

The value supplied for the LANG attribute is the two-letter ISO designation for that language, in this case "LA" for Latin.

Cross-Reference

You can find a full listing of ISO language designations at http://www.oasis-open.org/cover/iso639a.html.

Increase Efficiency

Good user interface design includes features that enhance the speed and efficiency at which the user can complete his or her tasks. This concept is no different in Web site design. Users want to be able to quickly locate and digest information. When they must interact with a Web site to fill out forms or travel between documents, that process should be intuitive and consistent.

The ACESSKEY **Attribute**

Moderately experienced computer users are likely to be familiar with the concept of the keyboard shortcut. That is, to access a specific program command or feature, you can activate it by using a combination of keyboard strokes. Windows users undoubtedly have used the Alt+F combination to access a program's File menu. Macintosh users have similar options when using the app key.

Web authors can provide similar keyboard access to elements in a Web page using the ACCESSKEY attribute. The following elements may be enhanced in this manner:

- A
- AREA
- BUTTON
- INPUT
- LABEL
- LEGEND
- TEXTAREA

No requirements are placed on the author as to what keys should be used, though common sense would dictate that some logical order be followed. An ordered list of three links could make use of the numbers 1, 2, and 3 for access keys as follows:

```
<OL>
<LI><A ACCESSKEY="1" HREF="page1.html" TITLE="Option 1">Option 1</A>
<LI><A ACCESSKEY="2" HREF="page2.html" TITLE="Option 2">Option 2</A>
<LI><A ACCESSKEY="3" HREF="page3.html" TITLE="Option 3">Option 3</A>
</OL>
```

The TABINDEX **Attribute**

If you've filled out online forms, you've probably run into those that don't behave as you expect when you try to tab between fields. Instead of moving horizontally across a set of fields, the cursor may tab down, or vice versa. Authors now have control over this function with the TABINDEX attribute. The same elements that accept the ACCESSKEY attribute can accept the TABINDEX attribute, so not only does tabbing apply to forms, but a user can readily tab through the links on a page to arrive at the one he or she wishes to follow.

The form seen here has a layout commonly used for collecting a user's name and address data. The HTML for this form is also provided.

```
<FORM>
First Name: <INPUT TYPE="text" NAME="FName" SIZE="20">
Last Name: <INPUT TYPE="text" NAME="LName" SIZE="20"><BR>
Street Address: <INPUT TYPE="text" NAME="address" SIZE="40"><BR>
City: <INPUT TYPE="text" NAME="city" SIZE="20">
State: <INPUT TYPE="text" NAME="state" SIZE="2">
Zip Code: <INPUT TYPE="text" NAME="zip" SIZE="10">
</FORM>
```

By adding a TABINDEX to each of these elements, you can direct a tab to move the user from First Name to Last Name, then down a line to Street Address, and logically through the rest of the form in the manner that U.S. residents are used to providing their address. The new markup is as follows:

```
<FORM>
First Name: <INPUT TYPE="text" NAME="FName" SIZE="20" TABINDEX="1">
Last Name: <INPUT TYPE="text" NAME="LName" SIZE="20" TABINDEX="2"><BR>
Street Address: <INPUT TYPE="text" NAME="address" SIZE="40" TABIN-
DEX="3"><BR>
City: <INPUT TYPE="text" NAME="city" SIZE="20" TABINDEX="4">
State: <INPUT TYPE="text" NAME="state" SIZE="2" TABINDEX="5">
Zip Code: <INPUT TYPE="text" NAME="zip" SIZE="10" TABINDEX="6">
</FORM>
```

Creating Accessible Forms

Access keys and tab indexes provide a good basis for creating accessible forms. However, there are several more forms-specific elements that provide even greater control.

SELECT Elements Using the OPTGROUP Element

There are times when a user must be guided through a complex list of response options when filling out forms. For instance, a troubleshooting request form may need to gather data about a user's computer processor.

With so many options in CPUs still in service, the list may grow confusingly long:

```
<SELECT name="CPU">
        <OPTION value="PII-400">Pentium II 400
        <OPTION value="PII-333">Pentium II 333
        <OPTION value="PII-300">Pentium II 300
        <OPTION value="PII-266">Pentium II 266
        <OPTION value="P266">Pentium 266
        <OPTION value="P233">Pentium 233
        <OPTION value="P166">Pentium 166
    </SELECT>
```

The new OPTGROUP element enables the logical grouping of options so that the user agent may render them in meaningful sections, perhaps as a hierarchical menu, or with some other semantic that reflects the added structure provided by the group. The HTML consists of the OPTGROUP element that contains a LABEL attribute:

```
<SELECT name="CPU">
        <OPTGROUP LABEL="Pentium II">
        <OPTION value="PII-400">Pentium II 400
        <OPTION value="PII-333">Pentium II 333
        <OPTION value="PII-300">Pentium II 300
        <OPTION value="PII-266">Pentium II 266
        </OPTGROUP>
        <OPTGROUP LABEL="Pentium">
        <OPTION value="P266">Pentium 266
        <OPTION value="P233">Pentium 233
        <OPTION value="P166">Pentium 166
        </OPTGROUP>
    </SELECT>
```

FIELDSET **and** LEGEND **Elements**

The FIELDSET element enables individual controls and labels to be combined into logically related groupings. A rather ubiquitous example is the controls that make up a user's mailing address. A typical form often includes the following:

```
Street Address: <INPUT type="text" name="address" size="40"><BR>
Address Line 2: <INPUT type="text" name="address2" size="40"><BR>
City: <INPUT TYPE="text" name="city" size="20">
State/Province: <INPUT type="text" name="state" size="10">
Zip/Postal Code: <INPUT type="text" name="zip" size="10">
```

These items are all logically a part of a grouping of address elements. To add that grouping to the HTML, the FIELDSET element is used as a container as follows:

```
<FIELDSET>
Street Address: <INPUT type="text" name="address" size="40"><BR>
Address Line 2: <INPUT type="text" name="address2" size="40"><BR>
City: <INPUT TYPE="text" name="city" size="20">
State/Province: <INPUT type="text" name="state" size="10">
Zip/Postal Code: <INPUT type="text" name="zip" size="10">
</FIELDSET>
```

The LEGEND element acts as a caption for the FIELDSET, and occurs immediately after the opening FIELDSET tag:

```
<FIELDSET>
<LEGEND>Your mailing address</LEGEND>
...fieldset content...
</FIELDSET>
```

Current popular browsers don't render the new LEGEND element; however, the concept is one that should be familiar to you from other user interfaces. Figure 5.4 shows an option dialog box for configuring the Eudora Pro e-mail program. The collection of check boxes bounded with a border and labeled Show Mailbox Columns is similar to the intent of the FIELDSET and LEGEND elements in HTML.

Figure 5.4   A logical grouping of form elements with a caption

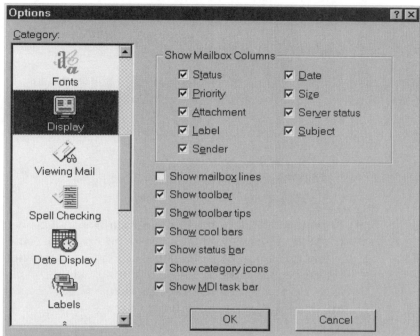

As with any text label, the legend improves accessibility for nonvisual renderings of the form. Additionally, it can provide reinforcement of the fieldset's purpose for those with visual user agents.

Frames and Tables

Most accessibility problems encountered with frames and tables arise from using these structural elements for aesthetic and presentational purposes. Before using these structures, review the reasons for including them. Do they simply arrange the content in a manner that is visually pleasing? Or do they provide a logical format for the data, such as an expense report presented in a table?

Frames

If you must use frames, the following three major requirements apply:

- **Always include the** NOFRAMES **element.** Typically, the content of the NOFRAMES element is a link to the unframed version of the site. This gives the user who cannot or does not want to view a site in frames easy access to your information. Remember, however, that your unframed content must also have navigation tools.

- The NOFRAMES element is presented within the FRAMESET declaration as follows:

```
<FRAMESET cols="50%, 50%">
        <FRAME src="content.html">
        <FRAME src="nav.html">
        <NOFRAMES>
        <P>Here is the <A href="content-noframes.html">
                non-frame based version of the document.</A>
        </NOFRAMES>
    </FRAMESET>
```

- **Provide a** LONGDESC **attribute in each frame.** This is especially important if the frame only contains visual information. This link can provide a means for a nonframes-capable user agent to provide information about the content of a frame before requiring the user to click through the NOF-RAMES content that is also supplied. The placement of the LONGDESC attribute is the same as it is for images:

```
<FRAME SRC="sidebar.html" LONGDESC="sidebar_desc.html">
```

- In the previous code, sidebar_desc.html might contain the following:

```
<BODY>
<P>This frame contains navigational links for use with the framed ver-
sion of this site. Visitors who wish, may view the <a href="stan-
dard.html">standard presentation</a> of this site.
</BODY>
```

- **Never use an image or object directly as the frame's source.** It is possible, though ill advised, to place an image file as a frame source.

```
<FRAME src="photo.jpg">
```

- This typically is seen when the only content of a frame is an image. The author may feel he or she has saved time or resources by not creating an additional HTML file that would serve as source for that frame. However, in doing this, accessibility barriers are raised: no ALT text is available for the image. The LONGDESC attribute is available, but you must remember that when used in a FRAME, the description is attached to the frame itself, not the framed content. If that content were to change, the

LONGDESC would no longer be applicable. Therefore, the solution is to create a traditional HTML file to use as the source of this frame, enabling appropriate ALT and LONGDESC attributes for the image itself.

Tables

Tables are some of the most abused elements in HTML. Early on, designers found that HTML as a structural language resulted in a rather fluid display across varying computer screens. One user might have a screen resolution of 640 ∞ 480 where another might be using 600 ∞ 800. These changes impacted the placement of elements on the page, as the browser fills its available space with document data. This is a great swing in mindset for those used to working on paper; once elements are printed on a sheet of paper, anyone who looks at that paper sees the same spatial relationships.

Captions and Summaries

In previous versions of HTML, authors had limited means to provide metainformation about a table. Any introductory explanations or captions weren't structurally linked to the table itself.

Tables now have two options for text-based descriptions: the CAPTION element and the SUMMARY attribute.

CAPTION is intended for a short description of the table's contents. It can be likened to the TITLE attribute for the overall document. Its placement comes immediately after the TABLE element (see Figure 5.5).

```
<TABLE>
<CAPTION>Today's Lunch Menu</CAPTION>
<TR>
<TH>Salad</TH>
<TH>Entrée</TH>
<TH>Dessert</TH>
</TR>
<TR>
<TD>Caesar</TD>
<TD>Chicken Divan</TD>
<TD>Chocolate Mousse</TD>
</TR>
</TABLE>
```

Figure 5.5   A table with a descriptive caption

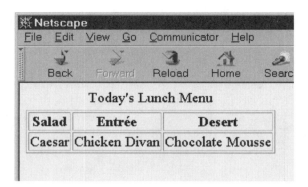

SUMMARY, on the other hand, provides a summary of the table's purpose and can also give clues to the structure of the table to assist in interpreting the output of voice- or Braille-based browsers. In very simplistic tables, such as the lunch menu sample, the caption and summary may contain similar text. However, when the table provides more than a simple list, the structural component of SUMMARY becomes increasingly important.

```
<TABLE SUMMARY="This table charts the dinner service menu features in
each of three classes of service on board Oceana Airlines flights:
Economy, Business, and First Class.">
<CAPTION>Oceana Airlines Dinner Service</CAPTION>
<TR>
<TH>Class of Service</TH>
<TH>Appetizer</TH>
<TH>Salad</TH>
<TH>Wine</TH>
<TH>Entrée</TH>
<TH>Dessert</TH>
</TR>
<TR>
<TD>Economy</TD>
<TD>Peanuts</TD>
<TD>none</TD>
<TD>none</TD>
<TD>complimentary soft drinks, liquor for purchase</TD>
<TD>Ham and Cheese sandwich</TD>
```

```
<TD>packaged cookies</TD>
</TR>
<TR>
<TD>Business</TD>
<TD>Shrimp Cocktail</TD>
<TD>Mixed Greens</TD>
<TD>Oceana label chardonnay or merlot</TD>
<TD>Chicken Satay or Beef Tips</TD>
<TD>New York style Cheesecake</TD>
</TR>
<TR>
<TD>First</TD>
<TD>Beluga Caviar</TD>
<TD>Select from Caesar, Salad Nicoise, or AntiPasto</TD>
<TD>Vive Clicquot Vintage Rèserve 1990</TD>
<TD>Salmon Basil Crème, Raspberry Balsamic Chicken, Boeuf En Daube</
TD>
<TD>Assorted Belgian truffles</TD>
</TR>
</TABLE>
```

Figure 5.6 shows this table in a visual user agent. Notice that the SUM-MARY attribute isn't displayed. However, screen readers and other assistive devices will likely read the summary first to provide the user with the appropriate context for the information to come.

II 5

Figure 5.6   **The** SUMMARY **attribute isn't rendered visually**

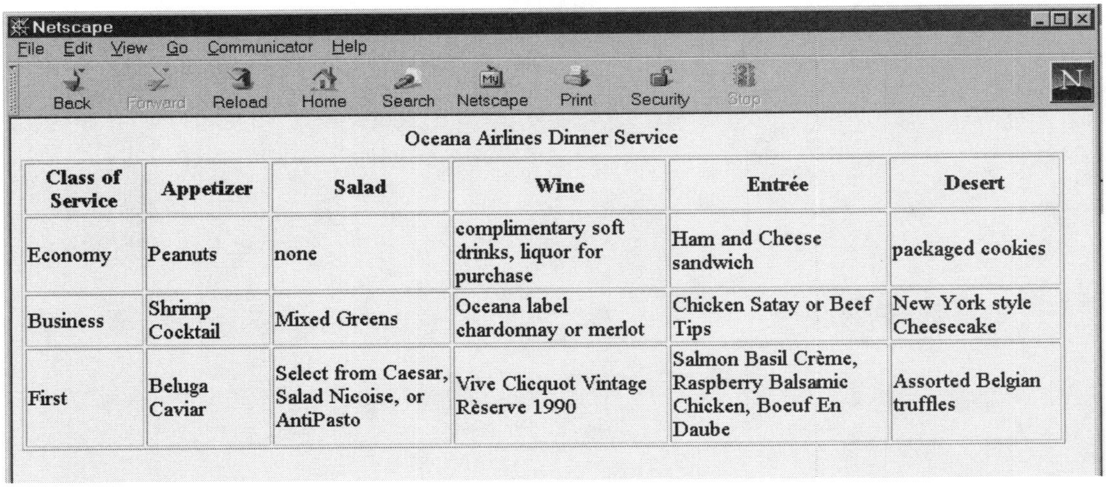

Scope and headers

Headers enable nonvisual user agents to retain the relational information conveyed by headings in a visual display. When viewing the visual rendering of Oceana's menu, the sighted user can repeatedly refer to the headings to reinforce the category for each dish presented.

To accomplish this, each header is given a unique ID attribute. Subsequent table cells then refer to the applicable header ID. Continuing with the Oceana dinner service example, the headers are written as follows:

```
<TABLE SUMMARY="This table charts the dinner service menu features in
each of three classes of service on board Oceana Airlines flights:
Economy, Business, and First Class.">
<CAPTION>Oceana Airlines Dinner Service</CAPTION>
<TR>
<TH id="h1">Class of Service</TH>
<TH id="h2">Appetizer</TH>
<TH id="h3">Salad</TH>
<TH id="h4">Wine</TH>
<TH id="h5">Entrée</TH>
<TH id="h6">Dessert</TH>
</TR>
```

```
<TR>
<TD headers="h1">Economy</TD>
<TD headers="h2">Peanuts</TD>
<TD headers="h3">None</TD>
<TD headers="h4">Complimentary Soft Drinks, Liquor for Purchase</TD>
<TD headers="h5">Ham and Cheese Sandwich</TD>
<TD headers="h6">Packaged Cookies</TD>
</TR>
<TR>
<TD headers="h1">Business</TD>
<TD headers="h2">Shrimp Cocktail</TD>
<TD headers="h3">Mixed Greens</TD>
<TD headers="h4">Oceana Label Chardonnay or Merlot</TD>
<TD headers="h5">Chicken Satay or Beef Tips</TD>
<TD headers="h6">New York Style Cheesecake</TD>
</TR>
<TR>
<TD headers="h1">First</TD>
<TD headers="h2">Beluga Caviar</TD>
<TD headers="h3">Select from Caesar, Salad Nicoise, or AntiPasto</TD>
<TD headers="h4">Vive Clicquot Vintage Rèserve 1990</TD>
<TD headers="h5">Salmon Basil Crème, Raspberry Balsamic Chicken, Boeuf
En Daube</TD>
<TD headers="h6">Assorted Belgian Truffles</TD>
</TR>
</TABLE>
```

This construction enables a screen reader to present the information as follows:

- Class of Service: Economy, Appetizer: Peanuts, Salad: None, Wine: Complimentary Soft Drinks, Liquor for Purchase, Entrée: Ham and Cheese Sandwich, Dessert: Packaged Cookies.

- Class of Service: Business, Appetizer: Shrimp Cocktail, Salad: Mixed Greens, Wine: Oceana Label Chardonnay or Merlot, Entrée: Chicken Satay or Beef Tips, Dessert: New York Style Cheesecake.

- Class of Service: First, Appetizer: Beluga Caviar, Salad: Select from Caesar, Salad Nicoise, or AntiPasto, Wine: Vive Clicquot Vintage Rèserve

1990, Entrée: Salmon Basil Crème, Raspberry Balsamic Chicken, Boeuf En Daube, Dessert: Assorted Belgian Truffles.

ABBR **attribute for a header**

When a header is relatively long, it can be cumbersome when repeated frequently as in the spoken example of our dinner service menu. To help avoid this, an abbreviation attribute can be applied to a header to shorten the presentation on subsequent appearances. For example, Class of Service is comparatively longer than the other headers. The abbreviation Class still imparts the pertinent information on subsequent readings. Therefore, the header element for Class of Service may be written as follows:

```
<TH id="h1" abbr="Class">Class of Service</TH>
```

When rendered by a screen reader, the first instance may be fully read as Class of Service, while subsequent readings would simply be voiced as Class.

The AXIS **Attribute**

Complex tables often present information that may be grouped across cells, columns, and rows. The tools provided in headers, IDs, and abbreviations can't fully address these needs. The AXIS attribute enables the relational grouping of data anywhere in a table.

The following table charts the protein, carbohydrate, and fat grams consumed by an individual at each of three meals over the course of two days (see Figure 5.7). Using headers, each entry for Grams of Protein is readily distinguishable from those for Grams of Fat no matter what type of browser is used. However, if a user wanted to know "How many grams of carbohydrate did I consume on Tuesday?" or "Did I eat more protein than carbohydrate for breakfast on Monday?," the related items are only grouped in that manner visually.

Figure 5.7   A complex table using axes and headers

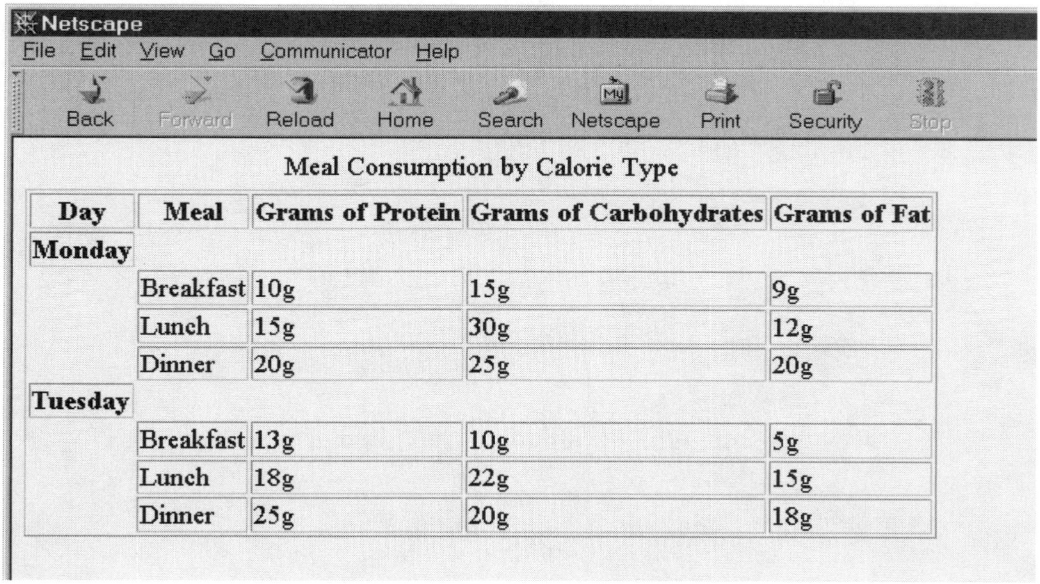

Meal Consumption by Calorie Type				
Day	**Meal**	**Grams of Protein**	**Grams of Carbohydrates**	**Grams of Fat**
Monday				
	Breakfast	10g	15g	9g
	Lunch	15g	30g	12g
	Dinner	20g	25g	20g
Tuesday				
	Breakfast	13g	10g	5g
	Lunch	18g	22g	15g
	Dinner	25g	20g	18g

In order to structure the table for all logical groupings, each unique category must have a unique ID. Thus, ID attributes are extended beyond the primary headings of Day, Meal, and each food type. New IDs are given to Monday and Tuesday, as well as each meal occurring on those days.

Each heading is then placed on an axis, or logical group. The day of the week is placed on the axis value "day." Meals are on the "meal" access, and each food type is on the "grams" axis.

Individual data cells are then aligned on the appropriate axis using the headers attribute. Protein grams consumed at Monday's breakfast would take the header values "h6" for Monday, "h8" for breakfast, and "h3" for grams of protein. The headers are written as a space-delimited list for a single headers attribute value such as:

```
<td headers="h6 h8 h3">10g</td>
```

The complete table markup appears as follows:

```html
<table border=1 summary="This table charts the consumption of protein,
carbohydrate and fat grams over three meals per day for two days.">
<caption>Meal Consumption by Calorie Type</caption>
<tr>
<th id="h1" axis="day">Day</th>
<th id="h2" axis="meal">Meal</th>
<th id="h3" axis="grams">Grams of Protein</th>
<th id="h4" axis="grams">Grams of Carbohydrates</th>
<th id="h5" axis="grams">Grams of Fat</th>
</tr>
<tr>
<th id="h6" axis="day">Monday</th>
<th></th>
<th></th>
<th></th>
<th></th>
</tr>
<tr>
<td></td>
<td id="h8" axis="meal">Breakfast</td>
<td headers="h6 h8 h3"> 10g</td>
<td headers="h6 h8 h4">15g</td>
<td headers="h6 h8 h5">9g</td>
</tr>
<tr>
<td></td>
<td id="h9" axis="meal">Lunch</td>
<td headers="h6 h9 h3">15g</td>
<td headers="h6 h9 h4">30g</td>
<td headers="h6 h9 h5">12g</td>
</tr>
<td></td>
<td id="h10" axis="meal">Dinner</td>
<td headers="h6 h10 h3">20g</td>
<td headers="h6 h10 h4">25g</td>
<td headers="h6 h10 h5">20g</td>
```

```
</tr>
<tr>
<th id="h7" axis="day">Tuesday</td>
<th></th>
<th></th>
<th></th>
<th></th>
</tr>
<tr>
<td></td>
<td id="h11" axis="meal">Breakfast</td>
<td headers="h7 h11 h3">13g</td>
<td headers="h7 h11 h4">10g</td>
<td headers="h7 h11 h5">5g</td>
</tr>
<td></td>
<td id="h12" axis="meal">Lunch</td>
<td headers="h7 h12 h3">18g</td>
<td headers="h7 h12 h4">22g</td>
<td headers="h7 h12 h5">15g</td>
</tr>
<td></td>
<td id="h13" axis="meal">Dinner</td>
<td headers="h7 h13 h3">25g</td>
<td headers="h7 h13 h4">20g</td>
<td headers="h7 h13 h5">18g</td>
</tr>
</table>
```

II 5

The HTML 4.0 recommendation does not require user agents to handle the axis data in any given fashion. However, implementers can program their agents to provide the semantic distinctions made available to the user through the careful structuring of the tables.

In answer to the user's question "How many grams of carbohydrates did I consume on Tuesday?," a user agent processing this table could report the following:

```
Day: Tuesday, Grams - Carbohydrates: 52g.
```

Locating all table cells aligned to both the headers "h7" for Tuesday and "h4" for carbohydrate grams produced this tally.

Summary

In this chapter you learned the following:

- The guiding principles of accessible Web design as presented by the Web Accessibility Initiative
- How to plan an accessibility retrofit of an existing Web site
- How to effectively plan for accessibility in a new Web site
- The importance of separating structure and presentation, and how that actually aids in presenting your site to all users
- What new features are included in HTML 4.0 that enhance accessibility
- Now that you've learned how to implement the HTML markup that will make your Web pages more accessible, it's important to test and validate your Web site. Chapter 6 shows you how to test your Web pages using Web accessibility validation tools and services.

References

World Wide Web Consortium. "W3C Techniques for Web Content Accessibility Guidelines." March 1999. Available from http://www.w3.org/TR/WAI-WEBCONTENT/wai-pageauth-tech.

World Wide Web Consortium. "W3C User Agent Guidelines." March 1999. Available from http://web1.w3.org/TR/WAI-USERAGENT/.

World Wide Web Consortium. "W3C Web Content Accessibility Guidelines 1.0." March 1999. Available from http://www.w3.org/TR/WAI-WEBCONTENT/.

Chapter 6

Testing and Validation

In This Chapter

- The importance of testing and validating
- Performing a Web site accessibility review
- Validation services
- Testing utilities and tools

This chapter is devoted to the subject of testing and validating Web sites. In this chapter, I lead you through a typical Web site accessibility testing process that I have found successful. I also discuss how to use the following validation services:

- Bobby
- W3C validation service for HTML and CSS2
- LIFT

Lastly, you will learn about the tools that can assist you during the testing process as well as utilities you can use to fix some of the more common Web accessibility barriers.

The Importance of Testing and Validating

Whether you're building a personal Web site or developing a corporate intranet, you'll never get anyone to visit your site if the quality of the site is poor. Almost without exception, users click the browser Stop button if they have to wait more than a minute or so for a site to load. Users take similar action when your Web site links are broken. Your site's validity comes into question.

The lack of accessibility plays into the Web site integrity factor too. If your site is not accessible, you are sure to lose millions of visitors — not just those who have disabilities, but also those who work with and otherwise support the accessibility community.

Take the time to test and validate your site before you post it. If you do so, you'll save yourself some embarrassment.

Note

Performing a Web site accessibility review is one stage of the Web site validation process. Concurrent with your accessibility review, you should perform a standard HTML validation review. To do this, I recommend the guidelines outlined in the *HTML 4 Bible* (Bryan Pfaffenberger and Alexis D. Gutzman, IDG Books Worldwide, 1998). These guidelines suggest that you test the most recent and previous versions of Netscape Navigator and Internet Explorer; Windows 3.1/95/98/ NT, Mac O/S, and UNIX platforms; and screen resolutions set to 640 ∞ 480, 800 ∞ 600, and 1024 ∞ 768.

Performing a Web Site Accessibility Review

Essentially you can perform one of the following two types of Web site accessibility reviews:

- A quick review using one or more HTML validation services and an accessibility validation service
- A thorough review, involving a combination of personal code review, services, tools, and user testing involving people with disabilities

Whether you perform a quick review or a thorough, top-to-bottom review of your site, it is important to verify that your site is accessible to people with disabilities. I recommend that you first focus on accessibility to the visually impaired and deaf because these are the two population segments most affected by inaccessible Web sites today.

Additionally, almost all accessibility validation services and tools cater to validation of HTML accessibility involving the visually impaired (including all strains of blindness) and the deaf. You can achieve a high degree of accessibility using this guideline.

Once you've succeeded in establishing complete accessibility for the visually impaired and deaf communities, review the material in Chapter 5 to learn additional steps you can take to expand the accessibility of your Web site to other disability communities.

In time, between guidelines, embedded publishing tool validation protocols, and validation services, you won't have to worry about prioritizing.

The Quick Accessibility Test

Quickly testing a site (or individual Web page) for accessibility is a simple, two-step process:

1. Test your HTML using the W3C's HTML validation service (`http://validator.w3.org/`) or Dave Raggett's HTML Tidy (`http://www.w3.org/People/Raggett/tidy/`) utility.

2. Test your Web site for accessibility using the Bobby (`http://www.cast.org/bobby`) or UsableNet's LIFT (`http://www.usablenet.com/index.htm`) accessibility validation services.

With the W3C's HTML validation service, you can easily check your HTML for coding syntax errors. To specify validation for HTML versions other than "strict" HTML 4.0, you must declare that version using the document type declaration at the start of your HTML document. For example to specify the HTML 4.0 Transitional DTD, you would include the following text:

```
<DOCTYPE HTML PUBLIC "-//W3C//DTD HTML 4.0 Transitional//EN"
"http://www.w3.org/TR/REC-html40/loose.dtd">
```

The W3C's validation service checks HTML that mandates accessibility (for example, the alt attribute is required syntax to the IMG element). For additional information about HTML versions and validation, please refer to W3C's HTML Web site at `http://www.w3.org/MarkUp/`.

Dave Raggett's HTML Tidy utility is easily the most useful authoring tool available. Not only will it clean up your HTML to insure that you are producing valid markup code, but it will offer you advice on how to make corrections to your code for accessibility. HTML Tidy uses the HTML 4.0 specification guidelines as its standard.

HTML Tidy originally existed as a UNIX command line tool. Today, you can configure MS Windows applications (for example, Word 97 and 2000) to use Tidy too. If you're a "purist" – that is, one who prefers to manually code your HTML using a favorite text editor, let me recommend that you consider downloading Chami.com's HTML-Kit, (`http://www.chami.com/html-kit/`) which includes embedded support for Tidy. The process for validating your pages is as simple as selecting the All or Tools tabs and clicking on the Validate icon (a little red icon of a hammer).

Remember, HTML 4.0 (the current W3C Recommendation is HTML 4.01) contains several enhancements to ensure accessibility. Validating your web site for HTML will significantly increase the accessibility of your web site to people with disabilities.

Once you've validated your HTML, use the Bobby or LIFT online web accessibility validation services. Both these services were designed to improve the accessibility and usability of your web site. And both use the W3C WAI Content Accessibility Guidelines as the basis for validation.

When performing a web site accessibility validation, try to get a sense of the major accessibility errors and/or the most frequently occurring accessibility errors. Once you've determined these, focus on correcting them first. Based on my personal experience, here are the most common web accessibility errors:

- Improper HTML 4.0 and/or CSS coding
- Lack of a site map
- Lack of text alternatives for images and image maps
- Structure elements used for presentation
- Poorly labeled forms and frames
- Poor presentation and writing style
- Use of PDF without providing alternative outputs or transformation services
- Javascript or other programming code
- Inaccessible multimedia (streaming video/audio)

CAST's Bobby analyzes your Web pages for accessibility to people with disabilities. Bobby also checks for browser compatibility. Note that the developers at CAST are continually working to keep Bobby up to date with the current release of the W3C guidelines, but this, like the guidelines themselves, is a continual work in progress. For information about how to use Bobby, see the next section, "Validation Services".

UsableNet.Com's LIFT Online validation service focuses primarily on the usability of your web site. LIFT looks for color usage, use of frames, browser compatibility, use of keywords, image size and text descriptions, and several other key usability metrics. For additional information about LIFT, see the next section, "Validation Services".

By the time this book is published, you will be able to go directly to WebABLE and use either Bobby or LIFT from the WebABLE homepage.

Note that there are other HTML and accessibility validation services. Many are discussed in sections that follow. However, to ensure the highest percentage of accessibility, it's important that you test both the HTML and the accessibility coding. If you use this process, you'll avoid most accessibility problems and become aware of other potential accessibility issues that require your consideration.

The Thorough Web Site Accessibility Review

You should perform a thorough Web site review in order to achieve the highest level of Web accessibility. To accomplish this, you *must* review the following areas:

- Images and image maps
- Links
- Text (paragraphs, horizontal lines, phrases, punctuation, and symbols)
- Color usage, coordination
- Lists and outlining
- Multimedia (audio and video)
- User-input forms
- Proper use of style sheets (CSS)
- Tables
- Frames
- Use of applets and scripts, blinking, or screen updating

If you are building a new Web site, review the preceding areas before you launch the site. If you operate an existing site, perform the review and make changes to the areas of accessibility that provide an immediate return on your investment. For example, if your site is designed using templates, a single template change will likely result in changes throughout your entire Web site.

There are tools available that automate or simplify the process of making accessibility changes. For example, a tool such as Michael Vorburger's ALTifier (http://www.vorburger.ch/projects/alt/) automatically examines your Web page for images and assists you in the creation of alternative text for the image. This kind of tool dramatically speeds up the process of making your site accessible to people with disabilities.

Additionally, ALTifier tests Web pages (graphical and text-only) using client applications against the HTML 3.2 standard.

Even though you should design to HTML 4.0 for better accessibility markup, also test all of the Web pages against the HTML 3.2 standard. Then test with and without graphics enabled using the following client browsers:

- Netscape Navigator (Version 2.0 and later)
- Internet Explorer (Version 3.0 and later)
- Lynx
- IBM HomePage Reader (self-voicing browser for the blind)
- pwWebSpeak (self-voicing browser for the blind)
- Opera

The primary objective of a Web site review is to ensure accessibility of your Web site for people with disabilities at the end-user level. The overall goal is to achieve accessibility now and through succeeding generations of your Web pages.

Color selections and contrasts are important to the readability of your web site. Improper color usage can make it very difficult for individuals with color deficiencies to use your site. To good references to consider when your designing your site are:

- Color Blind Design by Andrew Oakley
  http://www.delamare.unr.edu/cb/
- Diane Wilson's Site on Color Vision and Web Design
  http://www.lava.net/~dewilson/web/color.html

Additional Web site review and validation tools include:

- WHAT
  http://cmos-eng.rehab.uiuc.edu/what/
- A-Prompt
  http://aprompt.snow.utoronto.ca/

- pwWebSpeak
  `http://www.prodworks.com/pwwovw.htm`

WAI Recommended Testing Procedure

The Web Accessibility Initiative (WAI) Web Content Accessibility Guidelines recommend that you test your Web site by "validating your pages and assessing their accessibility with automated tools, manual tests, and other services." In short, they recommend you test your site by doing the following:

1. Using an accessibility validation tool or service
2. Testing with multiple browsers for a variety of conditions
3. Checking for spelling errors (assists people using screen readers or self-voicing browsers)

The WAI testing recommendations are based on a series of eight steps. You can review these steps within the W3C Web Content Accessibility Guidelines at `http://www.w3.org/TR/WD-WAI-PAGEAUTH/`.

Validation Services

Validation services for accessibility are still new in development. However, both Bobby and WebSAT perform so well that you can rely on them to capture most of the accessibility barriers that exist today.

Who Determines What's Valid?

HTML validation is a debatable topic. While the W3C provides tools to encourage HTML formal document markup — that is, markup that strictly conforms to the HTML 4.0 document type definition (DTD) — no one is policing Web sites to force conformance. Browsers and clients render content regardless of specification conformance. Most authoring tools support (or will begin to support) HTML 4.0 compliance, but again, no one except the Web page designer or site administrator can really enforce it.

Bobby

Bobby is an accessibility validation service that examines a single Web page or complete Web sites. Bobby validates your Web pages based on the Web Accessibility Initiative's Page Authoring Guidelines and it tests against a number of operating systems, browsers, special user agents, and HTML DTD levels.

Bobby is a Java-based program that runs as a standalone application on Sun Microsystems' Solaris (UNIX) operating system as well as Microsoft's Windows 95/98/NT platforms.

Web sites or pages that pass the Bobby review are encouraged to post the Bobby Approved icon (see Figure 6.1).

Figure 6.1 The Bobby Approved icon

Bobby is gaining worldwide acceptance as a legitimate validation service. The Center for Applied Special Technology (CAST), the creator of Bobby, cites the following statistics regarding Bobby's use:

- Each month an average of more than 3 million Web pages within several thousand Web sites are tested.
- More than 650 Web sites contain the Bobby Approved icon.
- Over 2500 sites are linked to or reference Bobby.
- Bobby has gained international acceptance.

Using Bobby to Validate Your Web Site

To review the accessibility of your Web site with Bobby, you have three options:

- Quickly review a single page against a single criterion.
- Review a single page for multiple criteria.
- Install and run Bobby as an application on your local system and review a complete Web site (local or remote).

Running Bobby on a Single Page with Single Evaluation Criterion

To validate a single Web page against one criterion, perform the following steps:

1. Go to the Bobby home page (`http://www.cast.crg/bobby/index.html`).
2. Type the URL of the Web page you want to validate in the URL entry field.
3. Select the appropriate evaluation criteria from the Evaluation Criteria pull-down menu.

4. Select whether you want the results of the validation returned in HTML Text-Only Output. This is optional. The default is to return the results in HTML, including graphics.

5. Click the Submit button. (Alternatively, you can press the Enter key.) The results are returned immediately.

Running Bobby on a Single Page with Multiple Evaluation Criteria

To validate a single Web page against multiple evaluation criteria, perform the following steps:

1. Go to the Advanced Bobby Options Web page (`http://www.cast.org/bobby/advanced.html`).

2. Type the URL of the Web page you want to validate in the URL entry field.

3. Select whether you want the results of the validation returned in Text-Only Output. This is optional. The default is to return the results in HTML.

4. Select the browser evaluation criteria. You can select from any of the following browser categories (identified with a check box):
   - America Online (Mac/Win v3.0, Win v2.5, Mac v2.6, Mac v2.7)
   - Microsoft Internet Explorer (v2.0, v3.0, v4.0)
   - Mosaic (v2.1.1)
   - Netscape Navigator (v1.1, v2.0, v3.0, v4.0)
   - Lynx (v2.7)
   - WebTV (v1.0)

5. Select the appropriate HTML Specifications check box. Your choices are HTML 2.0, HTML 3.2, and HTML 4.0.

6. Click the Submit button. (Alternatively, you can press the Enter key.) The results are returned immediately.

Running Bobby as a Local Application

You can download Bobby from the CAST home page and run it as an application on your personal computer. Whereas the Bobby Web site validation service only reviews one Web page at a time, the Bobby application can review a complete Web site and all its pages.

Bobby includes a complete reporting system, enabling you to specify the parameters of your validation. In addition, Bobby contains a pull-down menu that enables you to specify the extent of your URL analysis. Choices include the following:

- Don't follow links.
- Follow links in the URL's domain.
- Follow links in the URL's folder.
- Follow all links.

  You can also specify the number of hyperlink levels Bobby reviews. Once you've installed Bobby on your computer, run it as follows:

1. Start the application.
2. Type your Web site's URL in the URL address text entry field.
3. Click on the Go button. (Alternatively, you can press the Enter key.)
4. To display a report about a single URL, select the address in the Bobby report window and click the Report button. (Alternatively, you can press Ctrl+R.)
5. To display a complete Web site validation report, click the Summary button. (Alternatively, you can press Ctrl+F.)

The Bobby Reporting System

Bobby provides reporting status on six accessibility evaluation areas:

- Accessibility errors
- Recommended changes
- Accessibility questions
- Accessibility tips
- Browser compatibility errors
- Download time for Web page images

Making the accessibility error changes that Bobby recommends ensures that your Web site is minimally compatible for Web accessibility. Note that little blue helmet icons appear on your page if you have accessibility errors.

Recommended changes reduce noncompliance with the W3C Web Content Accessibility Guidelines. A significant number of accessibility recommendations and tips identify issues that may not be automatically recognized or categorized as more than potential problems.

Bobby then provides a list of accessibility tips that you should take under advisement. These tips are specifically produced to help you increase the level of accessibility for your Web site.

An invaluable part of the Bobby validation process is its reports regarding browser compatibility errors. You should attempt to develop your site so that it is compatible with the widest variety of browsers. Following the suggestions of the report helps you accomplish this goal.

Lastly, Bobby reports the download time of each image, applet, and object located on the page. Keeping download time to a minimum makes your visitors happy.

Webmasters and administrators are encouraged to make as many changes as are reasonably possible to their Web pages and sites. While not all pages will contain required changes, most pages will contain recommendations.

LIFT

LIFT Online is incredibly easy to use and understand. It's a form-based validation service (at the time of this writing, plans were in place to release it as an off-the-shelf product too) that sends you a complete report of your web site through e-mail. The beauty of the LIFT report is that it not only tells you what to fix and why, but it shows you exactly how to fix your code. A simple cut and paste will do the trick!

To use LIFT, go to their web site at `http://www.usablenet.com/index.htm` and then do the following:

- Enter the URL of your web site in the URL: form entry field
- Enter your e-mail address in the e-mail: form entry field
- Select your web site type in the Type: pull-down menu
- Click on the Go! Button

I find the LIFT report system much easier to read and understand compared to Bobby. On the other hand, LIFT does not currently provide the depth of accessibility review that Bobby provides. In it's first iteration, LIFT reviews your site for the following:

- HTML 4.0 compliance
- Portable colors for backgrounds, foregrounds, links and specific FONT elements
- New and visited links should be different
- No BLINK, MARQUEE and SPACER elements are used
- Headings are used to break up text heavy pages
- Sites using frames include meaningful NOFRAMES alternatives
- Browser-compatible specification of frame borders are used
- Web page downloading does not exceed 20 seconds for low bandwidth users

- Images include ALT text
- IMG includes the size of the image
- Images are not embedded within link labels
- Pages include a description and meaningful set of keywords
- Page titles are short and meaningful
- E-mail addresses are explicitly stated
- All external links work
- Web pages do not contain invisible elements

WebSAT

The National Institute on Standards and Technology (NIST) developed the Web Static Analyzer Tool (WebSAT) as part of its WebMetrics usability analysis tool suite. WebSAT is a very useful tool for Web page designers concerned with performing a thorough usability review of their Web page, including several areas consistent with accessibility practices.

The WebSAT service is not typically associated with Web accessibility reviews. However, a thorough Web site accessibility review requires verification processes of a Web site that are not limited to barriers involving people with disabilities. Rather, accessibility involves *all* user communities. WebSAT enables us to address a wider audience and ensures that clients get a well-rounded review.

WebSAT enables you to test up to five URLs at a time, which is a minor limitation. (Most tools, including the standalone version of the WebMetrics application, enable you to test complete Web sites.) You can download WebSAT and run it as a local application on UNIX, Windows 95/98, and NT platforms.

The NIST WebMetrics Tool Suite

The WebMetrics Tool Suite is a very useful Web site usability tool that assists Web site administrators, designers, and developers during the design stage. Best of all, it's free. WebMetrics consists of three tools.

WebSAT Checks HTML for site accessibility and usability.

WebCAT Enables a usability engineer to quickly construct and conduct category analysis of your Web site to ensure that site categories and the associated data within a category meets the expectations of the user.

WebVIP An online usability tool that enables you to conduct user testing of your Web site by helping you instrument the site for usability testing. Because it's a Web-based tool, WebVIP enables you to develop and construct remote usability tests.

WebSAT uses a subset of usability guidelines to perform a review. Using these guidelines (based on a variety of sources), WebSAT determines whether your site conforms to the various standards. Table 6.1 outlines the six categories that WebSAT reviews for.

Table 6.1   WebSAT Validation Categories

Category	Description
Accessibility	Inspects site for inclusion of alternative text and use of frames and color
Form Use	Tests site for form completion, properly labeled buttons, and the capability to reset or clear the form
Performance	Checks for image size, inclusion of height and width specifications, JPEG use, and banner size
Maintainability	Ensures that a site is optimally maintainable by checking for relative links and author information
Readability	Verifies that a site doesn't contain too many links on a page, and checks for use of marquees, scrolling and/or blinking text, and placement of horizontal lines

W3C Validation Services

As you might expect, the World Wide Web Consortium (W3C) supports validation services for both HTML and CSS. These are important services that should become part of your accessibility review and productivity tools.

The W3C HTML Validation Service

When you use the W3C's HTML validation service (http://validator.w3.org/), you verify that your HTML conforms to the W3C's HTML specification. The service is flexible and enables you to validate against current and past versions of HTML, as long as you specify the document type declaration at the beginning of your HTML document. To use the service, type in address of the Web site you are validating and click the Validate this URI button. Alternatively, you can press the Enter key on your keyboard.

The W3C HTML validation service enables you to specify a variety of reporting characteristics, including the following:

- Display Weblint results
- Show the source output
- Show the parse tree

- Run Weblint in pedantic mode
- Show an outline of the current Web page
- Exclude attributes from the parse tree

To perform a standard HTML validation review, it's generally not necessary to specify these additional attributes. However, Weblint is an excellent tool that aids in the review for Web site accessibility. I discuss Weblint in more detail later in this chapter.

The W3C CSS Validation Service

The W3C CSS validation service (`http://jigsaw.w3.org/css-validator/`) reviews your Web pages for proper Cascading Style Sheets (CSS) coding. CSS is quickly becoming an important accessibility aid because it enables you to separate structure from presentation features (such as color, spacing, and font type).

The W3C validator is flexible enough to check your CSS coding using three methods:

1. Specifying the URI
2. Copying the CSS code into a text entry box that is subsequently reviewed
3. Uploading the complete CSS coded file for validation

Other Validators

There are several validation services and validation applications you can use to test your Web pages. However, not many exist that formally test a page for accessibility characteristics.

In addition to the validation services previously mentioned, you can also use one of the validators in the following sections to check your HTML and accessibility features.

Weblint

Weblint is a freeware Web site validation tool created by Neil Bowers. It's fairly simple to install and tests for HTML syntax and style conformance.

As part of the validation process, Weblint checks that images include alt text, makes sure there is support for user and site configuration files, watches for overlapped or nested lists, reviews style, and tests for HTML that cannot be rendered by all browsers. All of these areas assist users with disabilities to more easily access a Web site.

Doctor HTML

Doctor HTML (http://www2.imagiware.com/RxHTML/) is a form-based Web page and site validation service. However, this is a fee-based site that enables guests to test up to four URLs. Doctor HTML performs analysis on the following:

- Spelling
- Image bandwidth and size, by name
- HTML structure
- Image analysis, including width, height, colors, and alternative text.
- Table structure
- Hyperlink verification including content type, size, and last modified.
- Form structure
- * Show HTML command hierarchy

Testing Utilities and Tools

Several tools and online services have been developed to help you improve the accessibility of your Web site. Most Web accessibility tools increase access by automatically generating text descriptions or renditions of your pages and images. Others provide functionality for enhancing tables, style sheets, forms, and programming scripts.

ALTifier

Michael Vorburger's Web ALTifier (http://www.vorburger.ch/projects/alt/) is a tool you can use to automate the task of including alternative (ALT attribute) text to images, image maps, objects, and frames on your Web site. ALTifier is comprised of a toolkit with various modules, including an autonomous enhancing filter and a graphical user interface (GUI) tool.

Essentially, ALTifier scans your Web site for HTML coding constructs like IMG or OBJECT, and checks for the presence of associated text descriptions — for example, ALT and TITLE. (The TITLE attribute is only checked for within frames and objects.)

After scanning your Web page, ALTifier then implements appropriate ALT text description using a protocol that guesses what the text should be based on document context (and other heuristics).

Figure 6.2 The ALTifier GUI front end

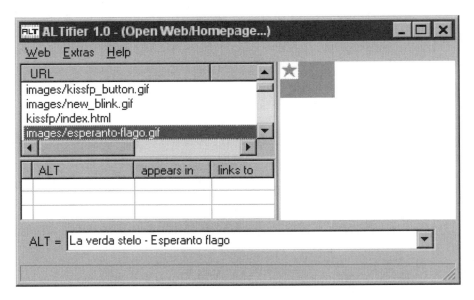

Figure 6.3 An example of the ALTifier's filtered HTML sample

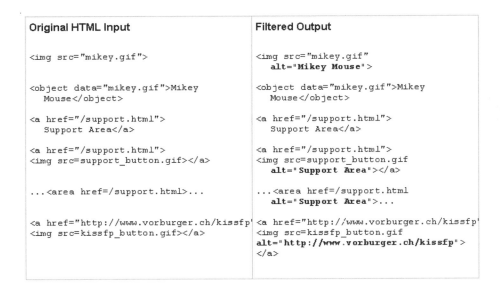

Lynx-It

People often wonder what a Web page "looks like" to a blind person. It's really not that complicated. All you need to do is view the page with a text browser like Lynx. If you don't have Lynx installed, then you'll appreciate the Lynx-It Web service (`http://www.slcc.edu/webguide/lynxit.html`) hosted by Salt Lake City Community College.

Lynx-It renders your Web pages as the Lynx browser would, but through a form-based service. Lynx-It also enables you to specify display attributes including numbering links and displaying a table of contents of links; showing visibly where image links point; and converting selection fields in forms from pop-up lists to radio buttons.

Lynx Viewer

If you don't have access to the Lynx text browser, Lynx Viewer (`http://www.delorie.com/web/lynxview.html`) is a quality Web-based service that renders your Web page as it would appear using Lynx.

Lynx Viewer is easy to use. Simply enter the URL for the Web page you want to render into the text entry field, and then click the View Page button. Alternatively, you can press the Enter key.

Note

Lynx Viewer removes the ability to tab through links, or follow them. Some found links are replaced by [INLINE]. Others in a list with "|" separators are only reported as a text stream, still separated by those "|".

NCSA's Text-Only Maker (TOM)

Once again, NCSA has demonstrated their continued support for Web accessibility through the creation of their Text-Only Maker (`http://lunch.ncsa.uiuc.edu/tom/tom.html`), better known as TOM. As its name implies, TOM enhances a Web site by including text descriptions for various HTML elements.

You can run TOM in two modes:

- **Replace** — Create a text-only page by replacing all the images with text
- **Add** — Include ALT text to graphics and images

Current TOM features include the following:

- Include ALT text for images or converting the image to text
- Convert server-side images to unnumbered lists
- Implement ALT text for client-side images

TOM requests input for each image that has no ALT text. A "clean" page, that is a page that does not contain images or contains images that are properly coded, results in a congratulatory message with no places to fix.

TOM gives the URL for each graphic without ALT text, but does not show the graphic. Therefore, you must view the original graphic in another application or user agent. Additionally, Web pages containing inconsistent use of ALT text may provide challenges when trying to determine which image (IMG) is being augmented with ALT text.

WebTV Viewer

Following are several good reasons why you should be sure to design your pages for WebTV:

- WebTV is the client of the future. Soon, more households will be connected to the Web through WebTV because of the technology convergence of the Web, cable, satellite, and television.
- WebTV design requires that you develop Web pages enhanced for keyboard input and navigation because the WebTV keyboard is an integral part of the WebTV system. Designing for keyboard input increases accessibility of the Web for the blind and others who rely on the keyboard to navigate the Web.
- WebTV renders Web pages based on television capabilities, which do not have the same quality screen resolution as your PC monitor.

The WebTV Viewer is a Web browser designed to display your Web site similar to the way the pages would be displayed through WebTV and your television. The WebTV Viewer home page describes some of the functionality of the WebTV Viewer:

- It shows what happens when pages are compressed horizontally to fit the display screen width of 544 pixels across.
- It shows how pages look after the WebTV browser enlarges HTML text to a large-sized Helvetica font to enhance readability on a TV screen.

- It supports the same HTML tags as a WebTV-based system, and most of the same multimedia technologies.

The WWW HTML Accessibility Tool (WHAT)

The WWW HTML Accessibility Tool (`http://cmos-eng.rehab.uiuc.edu/what/`) is a work in progress and, at the time of this writing, is still a prototype tool. However, with support from key accessibility organizations including Trace Research, NCSA, and the University of Illinois at Urbana-Champaign's Division of Rehabilitation-Education Services, this tool is certain to be an important instrument for Webmasters and Web site designers.

Like other tools, the initial focus of WHAT is to enhance a Web site by including text descriptions in appropriate places. WHAT currently enables checking and augmenting each image (IMG) with attribute values for ALT, LONGDESC, and TITLE, and/or adding a D-link to a separate URL describing the image.

The WHAT development team of Grace Tuan, Rick Langlois, Paul Hardin, and Jon Gunderson plan to strengthen WHAT by including the following features (as outlined on the WHAT home page):

- Convert inline font formatting to style sheets
- Assist in creating structured documents (converting font size and style specifications to H1, H2)
- Improve TABLE markup for accessibility
- Improve FRAME markup and add no-frames information
- Provide alternative representations for OBJECTS and APPLET
- Convert table positioning to CSS positioning

W3C's Tablin HTML Table Linearizer

An important new working group within the W3C's WAI program office is the Evaluation and Repair group. Recognizing the HTML coded tables can cause many accessibility problems, particularly for people who have visual disabilities, the ER group created the Tablin filter program. Essentially, Tablin transforms web pages that are coded with tables (whether for presentation for data purposes), into logical linear representations that are more easier read by screen readers and self-voicing browsers.

For additional information about Tablin, refer to `http://www.w3.org/WAI/Resources/Tablin/`.

Summary

In this chapter, you learned how to execute a Web site accessibility review. The review is a process accomplished by performing an in-depth code analysis along with using automated tools.

In the next chapter, I discuss the use of various browsers that support accessible viewing of Web pages.

Resources

Kasday, Len. *How to Check If a Web Page is Accessible to People with Disabilities Without Knowing HTML.* Available from `http://www.temple.edu/inst_disabilities/piat/webcheck/`.

Powell, Thomas. *The Complete Reference Web Design. 2000. Osborne/ McGraw-Hill publishers.*

Chapter 7

Browsing

In This Chapter

- Accessibility browsers
- GUI browsers
- Text browsers
- Telephony browsers

If the Web existed today as it did when Tim Berners-Lee first created it, accessibility would likely be nothing more than a minor inconvenience. Browsers at that time were strictly text-based, keyboard configurable, and highly usable by the blind community. Content was very accessible because little in the way of graphics was available.

NCSA's Mosaic changed all of that. Almost from the moment the graphical browser was introduced to the Web community, Web development surged into imagery and multimedia. Web designers created Web sites that drew upon the strengths of the graphical user interface and, subsequently, content became highly inaccessible resulting in the need to address several levels of access for people with disabilities.

In this chapter, you will see how a specific class of browsers has been developed for the disabilities community. You will also learn about the accessibility features of mainstream browsers, including Internet Explorer and Netscape Navigator. Finally, we'll look at a new generation of telephony browsers and services.

Accessibility Browsers

Once Mosaic became the standard and foundation for the first versions of graphical browsers, it was clear that several issues of accessibility would affect blind people who already were experiencing user interface problems introduced by operating systems with graphical user interfaces. Problems involving blind users led to the discovery of other barriers, including the lack of good font control and rendering for people with low vision, the implementation of browser keyboard shortcuts for people with physical disabilities (and blind people), and the lack of captioning for people who are deaf.

Initially, no accessibility features were available in the popular Web browsers Mosaic, Internet Explorer, and Netscape Navigator. Ultimately, this led to the development of specialized accessibility browsers by assistive technology vendors who focused primarily on Web accessibility for people with visual disabilities. Today, these browsers feature audible interpretation of Web pages, high contrast screen views, and screen magnification. Recently the addition of touch screens, voice recognition, and telephone browsing have increased the level of accessible browsing for people with disabilities.

But What About the Mac?

All the browsers described in the following sections were reviewed on the PC platform. Of these, most are compatible with Microsoft Windows 3.1/95/98 and NT, and many are compatible with the various UNIX windows operating systems, particularly Linux.

So what about the Mac? Frankly, the level of Web accessibility support at Apple is minimal at best. While several accessibility browsers are available on the PC platform, not one exists for the Mac.

But What About the Mac? (Continued)

Here's what I can tell you: Microsoft Internet Explorer and Netscape Navigator are available on the Mac OS platforms (System 7 and 8). However, this means that a blind person must use a windows screen reader in conjunction with either browser, because neither provides speech output. The only windows screen reader supported on the Mac platform is ALVA B.V.'s OutSpoken (`http://www.alva-bv.nl/`).

The MacLynx text browser (`http://www.lirmm.fr/~gutkneco/maclynx/maclynx_en.html`) is also available for the Mac and may be the most efficient Web browser available to users who are blind because the Mac has built-in speech (Speech Manager) functionality.

Opera Software is also planning to release a Mac-compatible version of their Opera Web browser, but it was not available at the time of this writing.

The following sections provide a comprehensive list of the accessibility browsers. Brief instructions on how to use the browsers as well as descriptions of their accessibility features are included.

pwWebSpeak Plus

The pwWebSpeak Plus (`http://www.prodworks.com/`) browser holds the distinction of being the first browser designed to enhance the Web experience for the blind and visually impaired. pwWebSpeak Plus, developed by The Productivity Works, Inc. (with support from DeWitt and Associates and Thomas Edison State College), is designed to interpret HTML and translate Web content into speech and a simplified visual display. pwWebSpeak Plus users can easily navigate through the structure of a document based on its contents, paragraphs, and sentences.

II

7

Figure 7.1   The pwWebSpeak Plus browser

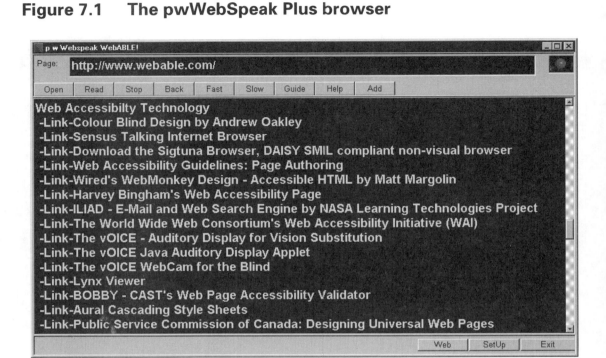

pwWebSpeak Plus supports a variety of output displays including synthetic speech, Braille output devices, and large print screen displays.

Accessibility Features

pwWebSpeak Plus includes the following accessibility features:

- Direct HTML interpretation using HTML Interpreter and Presentation Rule base.
- Audio and display synchronization for reading page elements.
- User-controlled font size to display large characters for low vision users.
- Simplified keyboard command structure for interacting with both pwWebSpeak and Web page content.
- Automatic recognition and audible identification of headings, links, forms, and other major page attributes. Users can review the page structure and associated links separate from the core page content. Users can audibly render one page construct (paragraph, word, or character) at a time.

- Complete access to tables, client-side images, and forms to enable full user interaction with applications and search engines.
- Support for a variety of internal and external speech engines. Reading speeds are adjustable.

Using pwWebSpeak Plus

pwWebSpeak Plus provides a standard set of command features that can be accessed through the Command menu or by entering definable keyboard commands. Blind users generally use the keyboard as their primary input device.

At any time you can press the Shift+F1 key combination to activate the audible command menu. The Command menu provides a list of nine main commands that you can select to perform all of the functions available in the browser. The nine main commands are as follows:

Open Page

On the Current page

Search Functions

Favorites

Save

Navigation Through Tables

Examine

Synthesizer Control

Control of Display

You can customize both Keyboard and Command menu commands. Table 7.1 provides a list of the most common keyboard commands.

Table 7.1   **Keyboard Commands for pwWebSpeak Plus**

Keyboard Commands	Function
Shift+F1	Activate Command menu.
F1	Help.

Keyboard Commands	Function
F2	Open new Web page. To open a local page, press F2 again.
F3	Read contents of entire Web page.
F4	Pause reading. Press F4 again to resume reading
F5	Select links list for links on current page. Press Enter to select link. User cursor keys to move between links.
F6	Identify current page element.
F7	Search the Web.
F8	Search current page.
F9	Identify current location. Press F9 again to read URL.
F10	Read summary of the Web page elements on current page.
Alt+A	Add current page to favorites list.
Alt+B	Return to previous Web page.
Alt+L	Slow down speech rate.
Alt+U	Set up pwWebSpeak.
Alt+X	Exit pwWebSpeak.
Alt+F10	Read current URL.
Shift+F2	Open Favorites List.
Shift+F3	Start playback of media stream.
Shift+F4	Stop playback of media stream.
Shift+F11	Increase character size.
Ctrl+F11	Decrease character size.
Down Arrow	Read next paragraph.
Up Arrow	Read previous paragraph.
Enter or Spacebar	Activate hyperlink.
Q	Stop reading.

pwWebSpeak Plus can also configured for use by people with low vision or people who are color blind. To increase the character size, press Shift+F11. To decrease character size, press Ctrl+F11. To adjust the browser foreground colors, press Shift+F12. To adjust the background colors, press Ctrl+F12.

Home Page Reader

IBM's Special Needs Systems group recently released their browser, Home Page Reader (http://www-3.ibm.com/able/hpr.html), which was developed exclusively for people who are blind and visually impaired. Using IBM's trademarked ViaVoice OutLoud text-to-speech speech synthesizer and Netscape Navigator, Home Page Reader provides accurate spoken output of Web information while at the same time providing a visual rendering of the data through the standard Navigator interface.

Home Page Reader was created by the IBM Japan Entry Systems Business Unit. In 1998, this group worked with the U.S.-based IBM Special Needs Systems organization to develop the English version for users of Microsoft Windows 95/98 and NT. Home Page Reader can also be used with screen readers for people who are visually impaired.

Home Page Reader is able to audibly render the complete contents of a Web site including graphics descriptions, text in column format, tables, and data input fields. No extra hardware synthesizer is needed. The user simply interacts with the computer using a basic numeric keypad in a Microsoft Windows environment. To start Home Page Reader, use the shortcut key combination Ctrl+Alt+H.

You can also navigate Web pages with frames by using the link keypad keys (1, 2, and 3). To open a frame, press + then keypad 2.

Table 7.2 lists the basic keypad commands.

Table 7.2   **Numeric Keypad Navigation Commands for Home Page Reader (Continued)**

Keypad Commands	Function
+ then .	Open URL
Num Lock and Enter	Exit browser
Enter	Stop reading
0	Read current page
1	Read previous link
2	Read current link
3	Read next link
4	Read previous item
5	Read current item

Keypad Commands	Function
6	Read next item
Num Lock	History list
/	Help
+ then /	Keys help
*	Settings
-	Bookmarks

Home Page Reader attempts to keep the user informed at all times. For example, after announcing that a URL is being loaded, the browser emits "beeps," which is an indicator that the page is still loading. If the page is not loaded within approximately 30 seconds, you are prompted to cancel the connection.

Accessibility Features

Home Page Reader includes several interesting and unique features that significantly enhance Web accessibility for people with disabilities. These features include the following:

- Speaks complete information about a Web site to the user, including the HTML 4.0 range of elements and attributes, tables, frames, forms, alternate text for images and image maps, and additional information including table summaries and captions.
- Navigates with the numeric keypad.
- Provides help mode for keys by pressing any Home Page Reader key combination, and retrieves a description of the key and how to use it.
- Contains a Fast Forward key that enables a user to quickly and efficiently scan Web pages for data. Also contains a Where am I command that provides the user with immediate navigational feedback, including information about the number and location of elements on a given Web page.
- Uses two voices — a male voice to read text, and a female voice to read link text — which makes it easier to comprehend site design and navigation.
- Installs easily using a voice-prompted installation and setup process.

BrookesTalk

The BrookesTalk Web browser (http://www.brookes.ac.uk/schools/cms/research/speech/btalk.htm) for the blind and visually impaired is function-key driven and includes a unique configurable text window so that visually impaired and sighted workers can work together. Developed by the School of Computing and Mathematical Sciences at Oxford Brookes University (http://www.brookes.ac.uk/) with help from the Royal National Institute for the Blind (http://www.rnib.org.uk), BrookesTalk provides Web page views using information retrieval and natural language processing techniques. BrookesTalk runs on Windows 95 and uses the Microsoft Speech Engine or Microsoft Speech API compatible software.

Users who find it difficult to remember keyboard functions will appreciate the built-in keyboard help feature: simultaneously press the Ctrl key and a function key, and BrookesTalk tells you that key's function.

BrookesTalk requires the Microsoft Speech Application Programming Interface 4.0 SDK or equivalent.

Accessibility Features

According to the BrookesTalk Web site, the browser includes the following accessibility features:

- A list of extracted keywords for each page
- Page summary including the number of words, links, headings, extracted keywords, and metakeywords
- Abstract comprising key sentences in the page (usually about 25 percent of the page)

  BrookesTalk contains two modes of Web page rendering:

- **Document** — Use this mode to read the whole document, list a page summary, and create a page abstract consisting of key sentences in the Web page.
- **Menu** — Use this mode to jump between the headings, links, and menus.

You can change the rendering mode depending on your need to review a particular portion of a Web page. BrookesTalk functionally breaks a document into separate chunks of information. A text chunk can be a whole document, the text between two headings, a paragraph, a sentence, or a word.

II

7

Table 7.3 describes many of the keyboard functions available to the user.

Table 7.3   **Keyboard Commands for BrookesTalk**

Keyboard Commands	Function
F1	Enter the URL and load page
F2	Search
F3	List page headings
F4	List page links
F5	List jumps
F6	List previous pages you visited
F7	Change browser settings
F9	List keywords
F11	Read page summary
F12	Read page abstract
Enter	Follow link
Spacebar	Pause and resume speech
Alt+F4	Exit browser
Alt+Up/Down Arrow	Increase/decrease speech speed
Alt+Left/Right Arrow	Increase/decrease speech volume
Ctrl+any key	Provide help on key function

Sensus Internet Browser

The Sensus Danish accessibility consulting firm (http://www.sensus.dk) developed the Sensus Internet Browser (http://www.sensus.dk/sib10uk.htm). The Sensus browser is supported on the Windows 95 operating system and works as an adjunct to Internet Explorer. Sensus renders Web content through its built-in software speech engine. Currently, Sensus supports English-language output. The goal of this Danish product is to support multilingual speech synthesizers and a dynamic language recognition module.

Version 1.0 of the Sensus Internet Browser is currently available. According to the Sensus Web site, plans for Version 2.0 include support for Braille, screen magnification, and multilingual speech synthesizers.

Figure 7.2   The Sensus Internet Browser

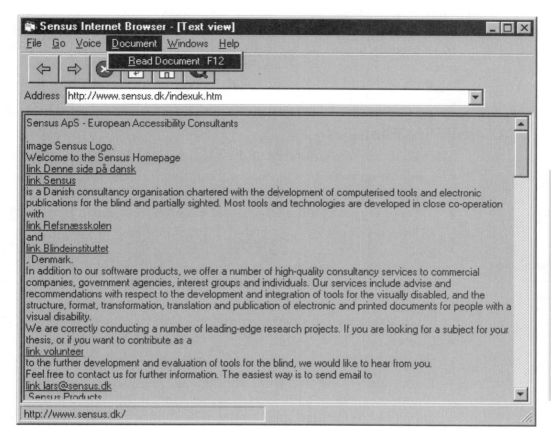

Accessibility Features

The Sensus Internet Browser was developed specifically to enhance the Web experience for people with visual disabilities through speech output. The Sensus Internet Browser, similar to other accessibility browsers, supports total keyboard control and navigation. This includes functions such as scrolling through the interactive links contained in a document or executing an interactive link. Navigation is facilitated through the keypad arrow keys and defined shortcut keys. As such, it includes the following accessibility features in addition to standard features already available through the Internet Explorer browser.

MultiWeb

MultiWeb is a browser designed by the Equity Access Research and Development Group at Deakin University in Australia. MultiWeb was developed to be used by people with a range of disabilities by providing different user interfaces for each type of disability. MultiWeb can be used without other adaptive software and includes a speech engine, text enlargement, and scanning for switch devices.

Accessibility Features

MultiWeb supports a number of useful accessibility features including:

- Interface support for a variety of input devices including the mouse, keyboard, switch device, and touch screen.
- Setup options including large print, highlight text, small buttons, speech synthesis, talking buttons, and a variety of colors and fonts.
- Inherited interface settings for the operating environment. For example, if a user selects the Switch Device Interface with the option of speech synthesis, then subsequent user actions within the Windows 95 environment also assume speech output.

MultiWeb consists of two separate programs:

- User Options (to select the interface and its associated options)
- MultiWeb (the actual browser)
- These options are accessible through the Start>Programs menu list in Windows 95/98. If you attempt to use MultiWeb, keep in mind that one of its primary features is to render Web content in large text. To do this, the browser takes up your entire PC screen. To get to other applications on your desktop, simply type Alt+Tab and select the application you want to use.

Sigtuna Browser

The Sigtuna Browser (http://www.jsrd.or.jp/dinf_us/software/browser.htm) is a Microsoft Windows 95, 98, and Windows NT program that reads and presents Web pages using synthetic speech and plays digital audio in both the Digital Audio-based Information System (Daisy) and RealAudio formats. The Sigtuna Browser is a version of the pwWebSpeak browser with the following distinctions:

1. Sigtuna does not include a speech synthesizer.

2. Sigtuna is supported by the Japanese Society for Rehabilitation of Persons with Disabilities (JSRPD).

3. Sigtuna is only available to:

   - Nonprofit organizations
   - Government groups within developing nations
   - Members of the DAISY Consortium
   - Government organizations in Japan

Sigtuna contains all the accessibility features of pwWebSpeak in addition to its ability to render digital talking book technology. Digital talking books should accelerate the possibility of providing people standard print information in a quality format for people who are visually impaired.

The Sigtuna Browser is available in both Japanese and English versions. The Japanese version requires the 95Reader software in order to provide support for voicing of IME keyboard entry. The English version requires you have a supported speech synthesizer including one of the following:

- 32-bit version of SoftVoice
- DECtalk Software Speech Synthesizer
- SAPI-compliant speech synthesizer such as Watson, Flextalk, or Microsoft's Speech Synthesizer.

To use a SAPI-compliant speech synthesizer, the synthesizer must register the voices it uses with the operating system, otherwise it is not supported in the current release.

VIP Browser

JBliss Imaging Systems' (http://www.jbliss.com/) VIP Browser is a Windows 95 talking Web browser (http://www.jbliss.com/SW_Products.html#VIP Browser) that optimizes visual displays for people with low vision using screen magnification. The browser includes voice output and contains a command directory sufficient enough to enable Web access for a blind person. The speech may also be turned on and off on demand.

The VIP Browser includes four types of screen magnified views:

- **Word wrap** — Displays text several lines at a time
- **Image view** — A split-screen view that displays standard lines of text in the upper view and magnified text in the lower view
- **Marquee** — Displays a single line of text at a time

- **RSVP (rapid serial visual presentation)** — Displays one word at a time

VIP Browser uses command menus that describe shortcut keys to help users operate the system. The browser includes an interactive learning mode that announces what keys do when users press them. A function key also summons a narrative help file.

Because VIP Browser displays only text and links by default, Web content rendering is fast and efficient, similar to rendering content in a text-mode browser. VIP Browser highlights and announces links, and provides users with the option to retrieve a list of all the links contained on the current page or individual links. Images are represented as icons that can be downloaded and magnified, if preferred.

Accessibility Features

Following is a brief list of the accessibility features available in the VIP Browser:

- Four screen magnification view types
- Speech output
- Shortcut key command menus
- Learning mode function key announcer
- Text browser–like interface that renders text and links while providing icons in place of Web page images

Enhancing Internet Access (EIA)

Sarsfield Solutions, with funding from the Australian Department of Communications and the Arts, developed the Enhancing Internet Access (EIA) touch screen browser (http://gippsnet.com.au/eiad/eiad.htm). EIA is the only browser specifically designed to enable easier Web browsing for people with cognitive disabilities. EIA has an embedded Awareness & Assessment Protocol (AAP) used to monitor response times and accuracy. Ultimately, data can be collected and used to improve the ability for a person with cognitive disabilities to use the Web.

The EIA browser is Windows 95–compatible and requires a sound card and touch screen monitor. Sarsfield Solutions provides a specialized home page for EIA browser users at http://gippsnet.com.au/eia/eia_01.htm.

Figure 7.3 shows the unique key characteristic in this browser — its touch screen interface. Using large onscreen buttons, this interface was designed primarily for public access in rehabilitation facilities, libraries, and private clinics.

Figure 7.3   The EIA touch screen browser

Accessibility Features

Because the EIA browser was specifically designed for users with disabilities, it includes a fairly large accessibility feature set including the following:

- Simplified "Push Button" browser.
- Easy and intuitive touch screen interface.
- Context-sensitive, onscreen keyboard that extends touch screen capability.
- Integrated interactive tutorial that provides interactive, step-by-step learning of the skills needed for Web browsing, particularly useful for individuals with cognitive disabilities.
- An interactive, integrated Web tutorial that leads on naturally from the Awareness Assessment Protocol. It finishes with the option to connect to the World Wide Web.

- Awareness & Assessment Protocol (AAP) tool for clinicians working with people with physical or cognitive disabilities that provides quantitative information about a clients abilities and likely difficulties in using the Web.

WebCite

WebCite (http://www.hear-it.com/webcite.html) provides World Wide Web accessibility for visually impaired and reading impaired people. At the time of this writing, this browser was under development by Compusult Limited of Canada and I was not able to try it out.

According to their Web page, WebCite enables users to browse the World Wide Web, send e-mail, download information, and control the WebCite application within a fully functional text-to-speech environment. Again, according to their Web page, WebCite is completely keyboard driven with limited commands for ease of use. It is also compatible with both Netscape Navigator and Internet Explorer.

Text Browsers

Text browsers are universally accessible to most people, regardless of ability. For the blind and visually impaired, they are much easier to use in conjunction with screen readers because of their text-based interface. For the most part, there is little about a text browser that is not accessible. Their only drawbacks involve lack of support for all of the Web protocols, including multimedia and programming.

Text browsers used by people with disabilities include those discussed in the following sections.

Lynx

Lynx (http://lynx.browser.org/) is a text browser for the World Wide Web. The most current version, Lynx 2.8.1, runs on VMS, UNIX, and Windows 95/98 and NT. Versions also exists for Windows 3.1/3.11, DOS, UNIX (various types), Max OS 7 and 8, and IBM's OS/2. See Subir Grewal's Web page for up-to-date information regarding Lynx operating system platform support.

Lynx, which was created at the University of Kansas, was the first text-based browser and still commands a large user base. It is completely accessible to screen readers for the blind and very often is their browser of choice. Current versions of Lynx include support for complex Web site HTML markup including forms, frames, and tables.

Figure 7.4   The Lynx text browser

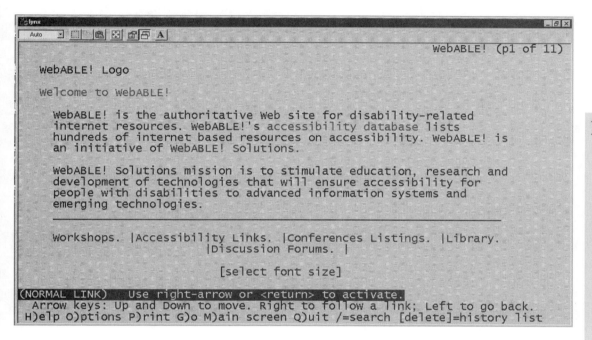

Net-Tamer

Net-Tamer, a product of Net-Tamer, Inc. (http://www.nettamer.net/tamer.html), is a DOS PPP dial-up access program that does not require a TSR packet driver. It browses the Web, receives and sends e-mail, receives and sends Usenet messages, uploads and downloads FTP files, Telnets to another Internet address, and checks the time and date. It is a both a robot and a navigator. It can also download and upload e-mail or Usenet messages on a timer. The program is speech friendly to users of talking programs for the visually impaired. Net-Tamer also runs on Palm Top handheld computers.

Emacs/W3

Emacs/W3 (`http://www.cs.inciana.edu/elisp/w3/docs.html`) is a full-featured Web browser. It supports all the available Web browsing functionality typically associated with the average GUI browser used on the Web today. Emacs/W3 includes complete support for forms, frames, tables and style sheets and runs on most operating systems, including almost any flavor of UNIX, Windows 95 and NT, AmigaDOS, OS/2, and VMS.

Emacs/W3 supports asynchronous connections, enabling users to browse numerous sites at the same time. The browser supports the Emacs mail and news reading packages, which enable easy sharing of information. Best of all, it works perfectly with Emacspeak (`http://cs.cornell.edu/home/raman/emacspeak/`), T.V. Raman's audio desktop for the blind. All of the browser's standard functionality as described below is completely accessible to Emacspeak. Subsequently, for accessibility purposes, this browser is completely accessible to users who enjoy speech-based browsing.

Features

- Asynchronous downloads enable multiple operations. This enables you to retrieve more than one file at once, or to continue browsing while downloading a large file in the background.
- All font/formatting control is now done through a default style sheet, which the user can override. This uses the W3C's Cascading Style Sheet specification.
- Tighter integration with standard tools in Emacs.
- Dired, a powerful directory browser, is used when visiting local or remote file systems, and enables actions on groups of files, instead of just one at a time as most Web browsers do.
- Easy-to-use preferences panel.
- Tool tips on hyperlinks and toolbar items when using XEmacs.
- Out of the box integration with Emacspeak, T.V. Raman's speech synthesizer package for Emacs and XEmacs.

GUI Browsers

From an accessibility standpoint, you could easily point a finger at graphical user interface (GUI) browsers and call them "the bad guys." After all, it was NCSA's Mosaic GUI browser that wreaked havoc on the visually impaired user community where access to the Web was concerned. Until Mosaic came on the scene, most Web surfers used the Lynx browser, which is text-based and accessible to the synthetic speech software used by blind people.

On the other hand, you could also say that the GUI browsers are responsible for raising the level of Web accessibility awareness. Because of the obvious lack of accessibility, users came out of their closets and clamored for support. However, this ruckus not only identified browser problems, but it also highlighted critical deficiencies in HTML and supporting tools.

The following sections describe GUI browsers used by people with disabilities, including people who are blind. Features that enhance accessibility or make the Web experience easier are noted.

Amaya

Amaya is the World Wide Web Consortium's own Web browser (http://www.w3.org/Amaya/) that also doubles as an authoring tool within the same window. Amaya serves an important function in the Web development arena because the W3C passes many of its new Web-based protocols through Amaya. For example, Amaya already provides support for CSS1 and MathML. Amaya is available for UNIX and Windows 95/NT.

Generally speaking, Amaya supports new W3C protocol development and therefore is a good GUI browser to have around. Its overall state of accessibility is fair at best. It does contain a Zoom feature in the View menu (Alt + or Alt - on keyboards) that enables users to increase the size of text and graphics as they appear on a Web page.

Amaya includes a WYSIWYG (What-You-See-Is-What-You-Get) editing interface. Assuming you have write access to a particular Web page, you can first browse a Web page and then edit it within the same window simply by single-clicking within the editing window. However, the authoring interface is not accessible to windows screen readers.

Cross-Reference

Refer to Chapter 8 for additional information about Amaya's editing interface.

Figure 7.5   Amaya Web browser

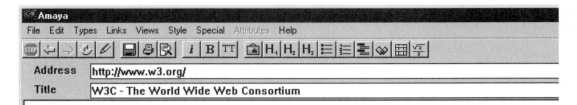

Internet Explorer

Microsoft Internet Explorer 5 (http://www.microsoft.com/enable/products/ie5.htm) is the latest version of Microsoft's Web browser. It includes a number of accessibility features that make accessing the Web easier for people with disabilities. Some of these features enable you to customize the appearance of Web pages to meet your preferences. IE 5 is available on Windows 95/98 or NT, as well as the Macintosh OS 8 operating system.

Microsoft Corporation has been involved in the accessibility business for several years, so it's no surprise that their Web browser, Internet Explorer, contains several accessibility features for people with disabilities. Much of their work is directly related to the efforts of Microsoft's Accessibility and Disabilities Group (http://www.microsoft.com/erable).

The current browser carries over all the features previously available in IE 4. Internet Explorer includes complete support for Microsoft Active Accessibility, the key underlying set of operating system–based protocols that enable an assistive technology (hardware or software) to work with other available installed software applications.

Similar to popular GUI browsers, including Netscape Navigator and Opera, Internet Explorer supports several core features that make Web browsing easier for people with disabilities. For example:

- Creating and installing style sheets
- Automatic completion of specified Web addresses (AutoComplete)
- Displaying ALT text for images and image maps
- Keyboard access and navigation
- Adjustable font colors, sizes, and styles
- Turn pictures, videos, and sounds off

Figure 7.6   Internet Explorer displays ALT text for images

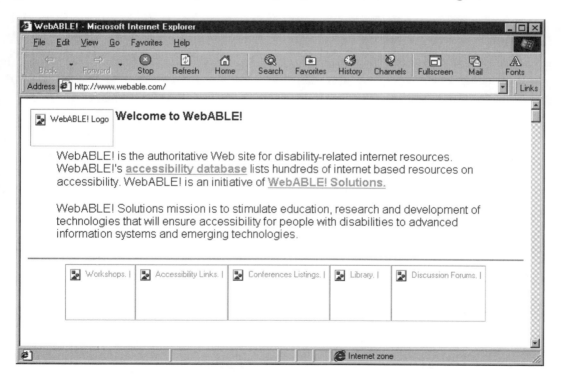

- In addition to the features previously listed, Internet Explorer also includes its own set of accessibility features.

Accessibility Features

Where accessibility is concerned, Internet Explorer 5 includes several features, many of which were introduced in previous versions of the browser. Essentially there are two ways to set accessibility features in Internet Explorer.

To specify font characteristics including color, size, and style, click the Tools menu, select Internet Options, and then click the Accessibility button (shown in Figure 7.7). The Accessibility property sheet is displayed, as shown in Figure 7.8. You can also use the Accessibility property sheet to specify the location of your own style sheet.

Figure 7.7   Internet Explorer's Accessibility button on the Internet Options dialog box

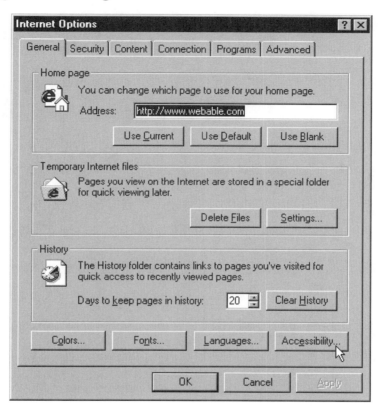

II

7

Figure 7.8   Use the Accessibility property sheet to set font characteristics and style sheets

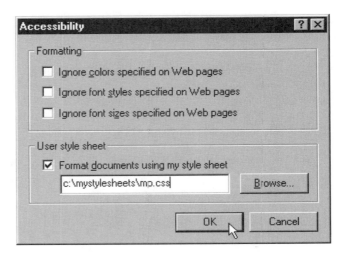

To specify expanded ALT text, set system caret focus, or turn image display off, use the Internet Options Advanced tab, as shown in Figure 7.9.

Figure 7.9 **Use the Advanced tab to turn image display off and support expanded ALT text**

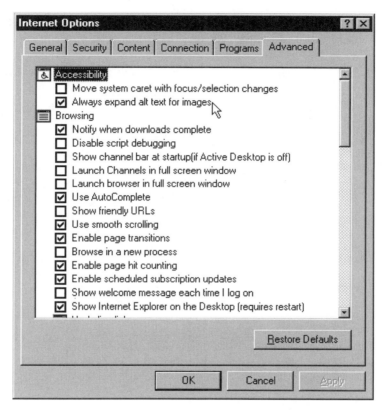

Additional options available through the Advanced tab are as follows:

- Fully displays the ALT text description when images aren't displayed.
- The system caret can be moved, enabling accessibility aids including screen readers and screen magnifiers the ability to track the area of the screen you are reading.
- Toolbars can be customized to enable you to make more text visible in the browser display. This is particularly useful for people with cognitive disabilities or people who prefer larger fonts.
- Optional sound cues inform you when your Web page begins and finishes loading.
- The ability to turn off or temporarily stop animations.

- The ability to disable smooth scrolling and other effects which can confuse screen reading utilities.
- Support for the High Contrast option in Windows.

Table 7.4 provides a short list of basic keyboard shortcuts for Internet Explorer 5 that are useful if you use a keyboard to navigate through the Web.

Table 7.4   Internet Explorer Keyboard Shortcut Commands

Keyboard Commands	Function
F1	Context-sensitive help for dialog box items
F5	Refresh current Web page
F6	Move forward between frames
Ctrl+F	Find on current Web page
Ctrl+E	Open Explorer Search
Ctrl+I	Open Favorites
Tab	Move between items on Web page
Shift+Tab	List keywords
Alt+Home	Go to home page
Alt+Right Arrow	Go to next Web page
Alt+Left Arrow	Go to previous Web page

Opera

The Opera browser (http://www.operasoftware.com) may be the big surprise entry where accessibility is concerned. Opera, developed by Opera Software AS of Oslo, Norway, was immediately heralded for its compactness and speed in addition to its support for accessibility. Opera includes full keyboard navigation, a customizable interface, and an easy-to-use zoom function that is particularly helpful for low vision users.

Beyond its accessibility features, Opera is a fully functional browser, providing all the features typically found in more popular commercial browsers such as Netscape Navigator and Microsoft Internet Explorer. The current public release of Opera, Version 3.5, contains substantial support for CSS1, HTML 3.2, Java, and a wide range of multimedia players. Opera is available for Windows 95/98 or NT. You can run Opera on Windows 3.1 and

3.11 systems as a native 16-bit application and, according to Opera's Web page, you can also use the browser on OS/2 systems.

Future development, as also reported by the company's Web site, includes additional support for OS/2, Mac OS, UNIX, and BeOS.

Accessibility Features

The Opera browser contains several key accessibility features:

- Screen reader–compatible hot lists.
- Fully accessible keyboard interface. Many keyboard functions are available via a single key (see).
- Ability to zoom both text and graphics, which is a valuable aid to low vision users.
- Selectable font size, style, and color.
- Sound feedback on browser startup, exit, clicking, completion of Web page loading, loading failure, and completion of transfer.

The beauty of the Opera browser is that its accessible features are an integral part of the interface. This can probably be attributed to Opera Software's developmental relationship with the Royal National Institute for the Blind, UK (http://www.rnib.org.uk). RNIB is heavily involved in the area of Web accessibility, as well as the Web Accessibility Initiative.

Opera includes a Zoom Document function on the main browser window. The Zoom function provides variable levels of screen magnification for users who find Web page text and/or graphics too small to view.

To use this function, click on the Zoom Document button, located (by default) in the lower-right corner of the browser window next to the URL address bar. Or, simply press the 0 key on your keyboard to zoom in by increments of 10 percent. Press the 7 key to zoom in by increments of 100 percent. The browser supports up to 1000 percent screen magnification.

Opera also includes settings to assist people who are blind and use windows screen readers with the Opera browser. For example, the default setting for the Opera's Lists menu item is to include icons on the menu bar. To remove those icons, do the following:

1. Select the Preferences menu item by typing Alt+P.
2. Select Hot List by pressing the H key.
3. Select Screen Reader Compatible Menus by pressing Tab four times.
4. Click OK.

Figure 7.10 Opera's screen reader–compatible menu support

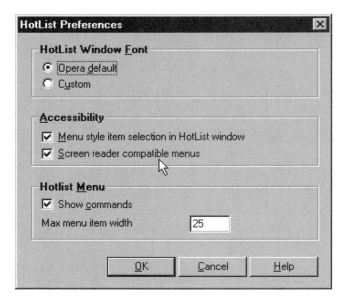

The hot list menu items now appear without the icons as displayed in Figure 7.11. This setting provides a more efficient way for screen readers to access menu items and read them to the user.

Figure 7.11   Opera's menus without icons

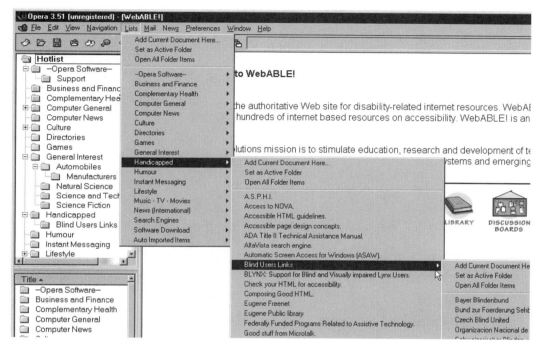

Screen reader users can jump between Web page elements and headers simply by pressing the keyboard keys D and E for elements, and W and S for headers.

Table 7.5 contains a list of some of Opera's basic keyboard shortcuts. For a complete list of all the keyboard shortcuts, select Keyboard from the Help menu. As an alternative, you can type Ctrl+B.

Table 7.5   Opera Keyboard Shortcuts

Keyboard Commands	Function
F1	Help
F2	Enter URL
F3	Search in current page
F7	Activate hot list
F8	Go to URL address bar

Table 7.5 **Opera Keyboard Shortcuts (Continued)**

Keyboard Commands	Function
Tab	Next element
Shift+Tab	Previous element
Ctrl+F3	View document source
Ctrl+B	Help on keyboard shortcuts
Alt+F3	View frame document source
Keyboard D	Jump forward between elements
Keyboard E	Jump backward between elements
Keyboard S	Jump forward between headers
Keyboard W	Jump backward between headers
Keyboard 0	Zoom in by 10 percent
Keyboard 6	Restore zoom to 100 percent (default setting)
Keyboard 7	Zoom in by 100 percent
Keyboard 9	Zoom out by 10 percent
Keyboard 8	Zoom out by 100 percent

Netscape Navigator

Netscape Navigator (http://www.netscape.com/) is considered by many Web surfers to be the most feature-rich browser available to users today. Navigator, the center of the Netscape Communicator application that includes an editing interface (Composer), address book, and the AOL Instant Messenger, is clearly a leader among browser and Web communication applications. And it's even better now that you can download it for free! Navigator is available on the all the major platforms including Windows 95/98 and NT, Mac OS, and UNIX.

However, on the accessibility development platform, Netscape has suffered from the perception that Navigator is not accessible, particularly to people with visual disabilities. In part, this is due to past market demand and development priorities for the company. However, it should be noted that Netscape developers have worked with assistive technology companies who produce windows screen readers for the blind to make Navigator more accessible. Today, many users with visual disabilities use Navigator along with their screen readers, surfing the Web just as easily and efficiently as users with specialized accessibility browsers.

Perhaps Netscape's most significant contribution to accessibility is release of the browser source code to the public through Mozilla.org (http://www.mozilla.org/). Now Web developers who create and support technologies for people with disabilities can design their applications to work smoothly with the browser. All things considered, this is a big step forward toward building an accessible Web.

Accessibility Features

Navigator includes support for several constructs that make Web browsing easier for people with disabilities. These include the following:

- Keyboard shortcuts (refer to Table 7.6)
- Style sheet support
- Customizable font settings for style, size, and color
- Automatic unloading of images
- Display of ALT text
- To increase the font size within Navigator, type Ctrl+] or perform the following steps:

1. Click the View menu.
2. Select the Increase Font menu item.

    Figure 7.12 illustrates how to use this setting.

II

7

Figure 7.12   Select Increase Font in Navigator's View menu to increase font size

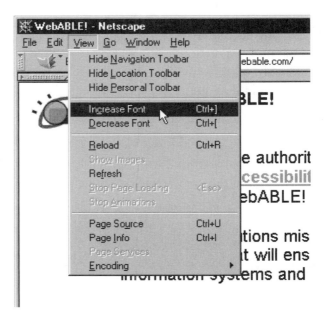

To unload images or enable style sheets (see Figure 7.13), perform the following steps:

1. Click the Edit menu.
2. Select the Preferences menu item.
3. Select Advanced from the Category window.
4. Click the Automatically load images check box to remove the check from the box.
5. Click the Enable style sheets check box to insert the check and turn on style sheet support.

Figure 7.13 Use Preferences settings to load images or enable style sheets in Navigator

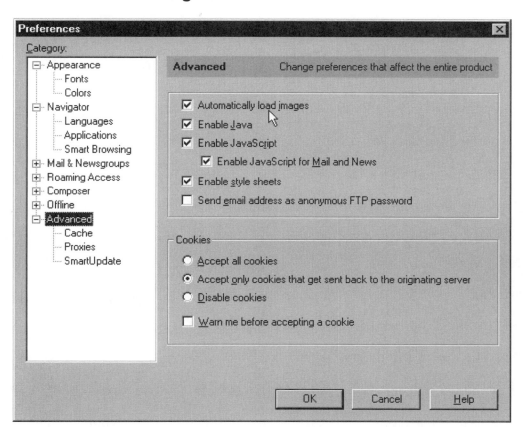

Table 7.6 is a compilation of some Navigator keyboard shortcuts available to the Windows 95/98 or NT user. To find the keyboard shortcuts for Mac OS or UNIX, refer to Netscape's keyboard shortcut help page at `http://help.netscape.com/products/client/communicator/qrc/nvshrtpg.htm`.

Table 7.6   Netscape Navigator Keyboard Shortcut Commands

Keyboard Commands	Function
F1	Context-sensitive help for dialog box items
F5	Refresh current Web page
F6	Move forward between frames
Ctrl+D	Add bookmark
Ctrl+E	Open Explorer Search
Ctrl+F	Find on current Web page
Ctrl+I	Open Favorites
Ctrl+O	Open page (URL or local file)
Ctrl+Q	Exit Navigator
Tab	Move between items on Web page
Shift+Tab	List keywords
Alt+Home	Go to home page
Alt+Right Arrow	Go to next Web page
Alt+Left Arrow	Go to previous Web page

Telephony Systems

Telephony browsers, systems, and services are beginning to enter the market. In most cases, the browsers are not specifically designed for accessibility, but many do contain features that make it easier for some people with disabilities to use. There are, of course, many other features about these browsers that make them attractive to all users.

Two interesting entrants into this field are the Microsoft Cordless Phone System and the SpeecHTML Web service.

Brief descriptions of each system follows. Keep in mind that these are not the only telephony browsers or Web-based telephony services available today — there are several others. However, characteristics of each interface employ technology that is usable by people with disabilities. If this is the wave of the future, focusing on the value of these characteristics and enhancing them gets us closer to the goal of universal accessiblity.

Microsoft Phone

The Microsoft Cordless Phone System is a sophisticated telephone system that works in conjunction with your personal computer. The phone components include a cordless phone, a base station, and Call Manager software you install on your PC. The base station connects to your PC through the serial port connection and interacts with the telephone through radio waves.

The fact that this system includes specific accessibility features again highlights Microsoft's commitment to accessibility for people with disabilities. Of course, some of the accessible features are standard to many standard telephone systems based on required engineering standards for telephones.

Accessibility Features

The Microsoft Phone System includes accessibility features for users who have hearing or visual disabilities. Additional accessibility features for people with physical disabilities are also included.

The following key accessibility features of the Microsoft Phone System enhance usability for a person with a disability:

- Phone device is hearing-aid compatible
- Voice-oriented interface for all users
- Visual indicator lights for all key signals
- TTY windows through the Call Manager Software for people who are deaf or have hearing disabilities (on the PC)
- User documentation is formatted for use by screen readers and available on CD-ROM

SpeecHTML

If the commerce wave of the Web is service (and I believe it is), SpeecHTML (http://www.speechtml.com/) is one service that will immediately have a positive impact on accessibility for people with disabilities. The current level of accessibility is thin, but as we move toward the information appliance age, accessibility enhancements for people with disabilities will become an important requirement.

For example, the SpeecHTML site is designed with frames with no content contained within the NOFRAMES element, and it does not include ALT text for images. As previously noted, it can be difficult to navigate a Web site with screen readers or text browsers that do not support frames and rely on alternative text for graphics. Because information appliances such as SpeecHTML are likely to generate revenue by through public service, the importance of being accessible to everyone is critical.

How does the service work today? SpeecHTML is a subscription-based gateway between a telephone client and your Web site. The Web site owner pays a monthly subscription fee and is assigned a telephone number for the Web site. Individuals dial the number, which then connects them to the SpeecHTML gateway. The service accesses your Web site data, converts it to speech using a text-to-speech converter, and renders it through the telephone client to the user.

If your Web site includes links (and likely it does), each link is presented as a series of choices to the caller. Using a speech recognition application, the caller then vocalizes a selection and the link is activated and read back to the caller. If the Web site includes forms, the gateway vocalizes the selections and the speech recognition engine again acts on the response.

Clearly, telephony technology presents usability and accessibility issues that require attention. The HTML coding will have to be flexible for voice-based systems. Web sites functioning as help desks and general information services today particularly will be affected. What about access for people who are deaf and use telecommunications devices for the deaf (TDD)? Is the speech recognition engine good enough to understand the voice of a person with a speech disability? These are just a few of the questions and requirements that remain on the figurative development plate.

Summary

In this chapter, I reviewed different types of Web browsers — primarily for the Windows platform — and examined how they support people with disabilities. There are four categories of browsers that support various levels of accessibility: Accessibility, GUI, Text, and Telephony.

Clearly, the accessibility Web browsers have been designed to enhance the Web experience for a variety of people with disabilities. The accessibility browsers support features including spoken output, screen magnification, and touch screen access.

Some of the mainstream browsers, such as Internet Explorer and Opera, do contain a considerable degree of accessibility and are often accessible with screen reader software for the blind. Preference settings, style sheet support, and large text support are just a few of the features available in these browsers.

I've now covered how to create accessible Web sites, the testing and validation process, and tools and browsers that support accessible Web page rendering. In the next chapter, you'll learn about authoring tools and their level of support for accessibility.

References

Bill Perry's Emacs/W3 Web site.

DAISY Consortium Web site.

EIA Web site.

Emacspeak Web site. http://cs.cornell.edu/home/raman/emacspeak/.

IBM Special Needs Systems Web site. http://www-3.ibm.com/.

Jbliss Imaging Systems Web site. http://www.jbliss.com/.

Microsoft Accessibility and Disabilities Web site. http://www.microsoft.com/enable/.

Microsoft Corporation Web site. http://www.microsoft.com/.

Microsoft Phone Web site.

Mozilla.Org Web site. http://www.mozilla.org/.

MultiWeb Web site.

Netscape Communications Corporation Web site. http://www.netscape.com/.

Net-Tamer, Inc. Web site. http://www.nettamer.net/.

Opera Software Web site. http://www.operasoftware.com/.

Oxford Brookes University, School of Computing and Mathematical Sciences Web site. http://www.brookes.ac.uk/.

The Productivity Works, Inc. Web site. http://www.prodworks.com/.

Royal National Institute for the Blind Web site. http://www.rnib.org.uk/.

Sensus Web site. http://www.sensus.dk/.

Sigtuna Web site. http://www.jsrd.or.jp/dinf_us/software/browser.htm.

Subir Grewal's Extremely Lynx Web site.

II

7

W3C's Amaya Web site. http://www.w3.org/Amaya/.

WebCite Web site. http://www.hear-it.com/webcite.html.

8

Chapter 8

Publishing Tools

In This Chapter

- Why worry about tools?
- Text editors
- Visual editors
- Converters

Chapter 8 introduces you to the types of publishing tools you can use to create HTML documents, with focus on how the tool can help (or hinder) the production of accessible Web sites. The chapter examines three types of Web authoring tools: text editors, visual editors, and converters. Because Web development tools are continually upgraded, this chapter examines how the tools support accessible Web site design, rather than covering all available features.

Why Worry About Tools?

The World Wide Web is an open community of hypertext documents, servers, and browsers. Deviations from standard HTML can serve as roadblocks to the openness of this medium. Writing HTML documents requires only a text editor, such as Notepad in Windows, SimpleText on the Macintosh, or vi for UNIX. However, using such a non-Web-specific text editor means that you must know (and thus type) all the HTML tags that tell the Web client how to display the Web document.

As Web documents have become more complex, software companies have provided increasingly robust tools designed specifically for Web development. These tools may provide wizards for HTML tag completion, tag palettes, or a graphical interface that serve as a buffer between the author and raw HTML. Other tools automatically convert documents from a native file format, such as WordPerfect, Excel, or PageMaker, into raw HTML code.

Many authors are unaware of accessibility issues and create nonaccessible sites by default, not design. Therefore, it is important that a tool help Web authors develop accessible Web sites. Does the tool make it easy — through prompts, automated tools, checking and repair functions, help files, or alerts — to generate HTML that is compliant and accessible? And what is the quality of the code produced by the tool?

Other questions that need to be addressed with regard to Web authoring tools are as follows:

- Does the tool support all accessible content requirements and recommendations for HTML 4.0, such as using the ALT (alternative text) attribute for graphics and other objects?
- Does the tool encourage the Web author to create accessible HTML?
- Does the tool identify nonaccessible or nonstandard HTML?
- Does the graphical or conversion tool provide easy access to and easy edits of the HTML generated by the tool?

A-Prompt Toolkit for HTML Editors

In an effort to assist Web content publishers and vendors who develop Web publishing software, the Assistive Technology Resource Centre (ATRC) at the University of Toronto in collaboration with Trace Research and Development at the University of Wisconsin at Madison have developed a new software toolkit called A-Prompt. A-Prompt is a software module designed to be embedded within an HTML editor (although you can run it standalone) that prompts Web site creators for accessibility HTML code enhancements.

A-Prompt Toolkit for HTML Editors (Continued)

The A-Prompt toolkit is designed to increase Web accessibility and accessibility awareness by assisting Web page creators during the content production process. A-Prompt includes a validator that reviews elements for accessibility. If the validator determines that the HTML code is not accessible, the A-Prompt Wizard displays a warning message and a Repair box. Follow the instructions within the Repair box and you're on your way toward making your Web site more accessible.

Figure 8.1　A-Prompt standalone module

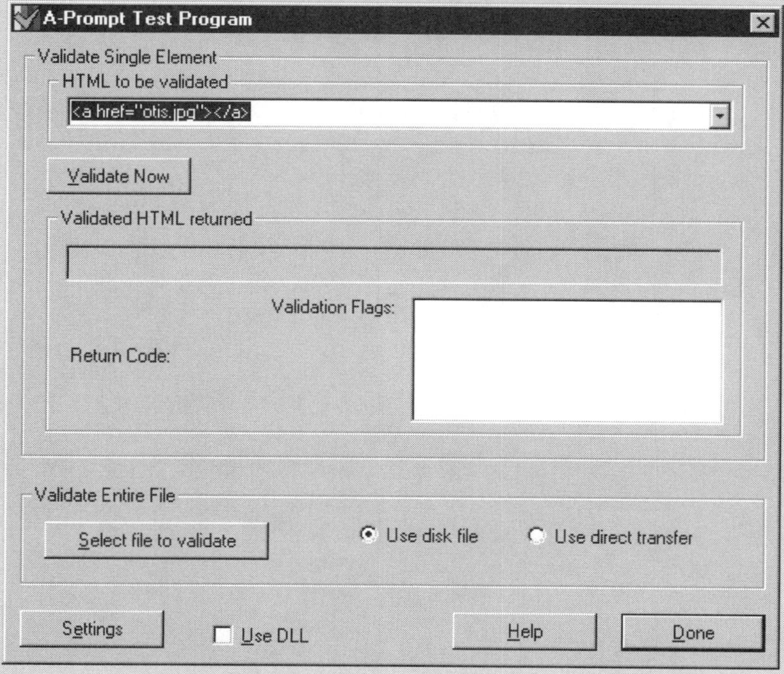

A-Prompt uses the WAI guidelines as the basis for conducting validation and repair. Currently, the beta version checks and provides repairs for 22 accessibility guidelines. Additionally, the A-Prompt toolkit uses an Alt text registry to facilitate the entry of Alt text.

You can download the A-Prompt toolkit and associated documentation from http://aprompt.snow.utoronto.ca/.

A-Prompt Toolkit for HTML Editors (Continued)

Figure 8.2   A-Prompt settings for repairs

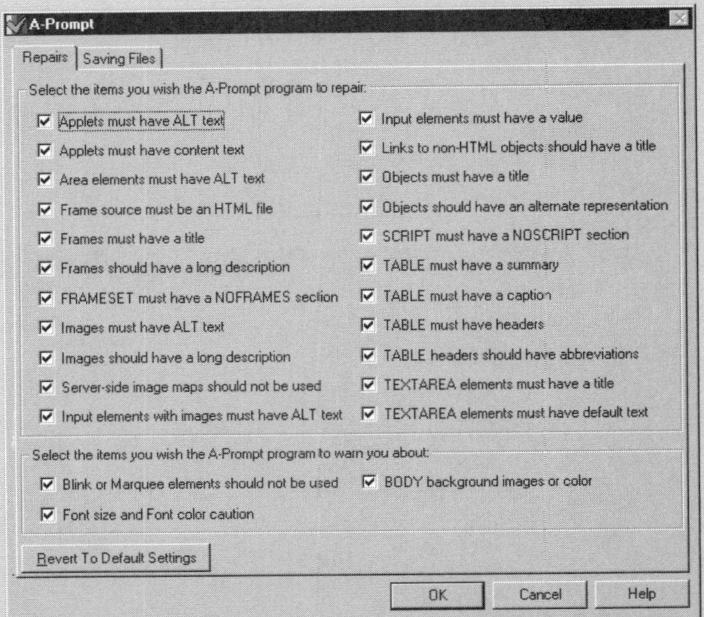

Figure 8.3   A-Prompt validation window

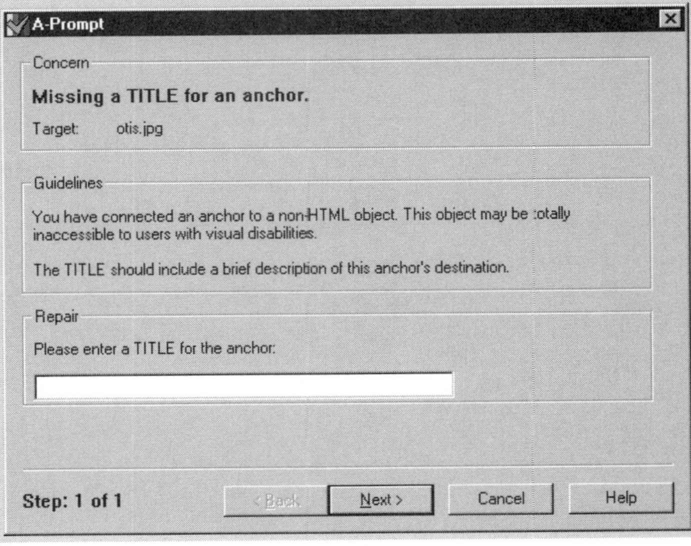

A-Prompt Toolkit for HTML Editors (Continued)

Figure 8.4   A-Prompt validation window with corrected HTML

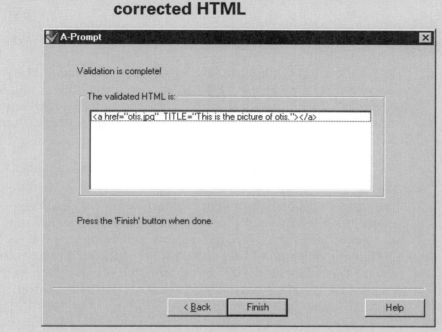

Text Editors

An HTML document is, at its heart, an ASCII document. Therefore, any word processing program that can save text as ASCII could be considered a text editor. For the purposes of this chapter, *text editor* refers to a Web development tool that is fundamentally a text-based document editor. It may have shortcuts, wizards, or palettes to provide easy access to some or all of the HTML tags, but it is primarily a plain text editor.

Two popular and award-winning text editors are BBEdit for the Macintosh and HomeSite for Windows. Each program is written only for one platform, and there are functional similarities (some of the differences reflect the interface common to each operating system).

Both text editors provide a Web publisher with ultimate control over the content and appearance of the code in the HTML document. Each also assumes that the Web author has a basic knowledge of HTML, such as which tags to use to mark content (although the author is not required to type the tags). Remember, HTML is a document markup language. It does not yet provide the precise presentation control that word processing or desktop publishing programs might. One reason for this difference in orientation is the extreme variability in output devices for Web pages.

BBEdit

BBEdit, created by BareBones Software (http://www.bbedit.com/), is a high-performance text and HTML editor for the Macintosh. As the tool has evolved, its support for HTML authoring has become more robust and comprehensive. However, it can be used for any text editing task, including writing programming code. It is the most highly acclaimed HTML text editor for the Macintosh.

After you launch BBEdit, the program opens a plain document in an editing window and the HTML Tools palette. For authors who prefer drop-down menus, the HTML tags are also accessible off the Markup menu. All HTML 4.0 tags and attributes are supported in these two environments, making reliance on memory a thing of the past.

The first line of code in a valid HTML document refers to its SGML document type definition (DTD), which declares structure and the elements supported. HTML 4.0 specifies three DTDs, so you must include one of the three DTDs in your documents.

- The HTML 4.0 Strict DTD includes all elements and attributes that have not been deprecated or do not appear in frameset documents. For documents that use this DTD, use the following document type declaration:

```
<!DOCTYPE HTML PUBLIC "-//W3C//DTD HTML 4.0//EN" "http://www.w3.org/
TR/REC-html140/strict.dtd">
```

- The HTML 4.0 Transitional DTD includes everything in the Strict DTD plus deprecated elements and attributes (most of which concern appearance, such as font and bold). For these documents, use the following declaration:

```
<!DOCTYPE HTML PUBLIC "-//W3C//DTD HTML 4.0 Transitional//EN" "http://
www.w3.org/TR/REC-html140/loose.dtd">
```

- The HTML 4.0 Frameset DTD includes everything in the Transitional DTD plus frames. For documents that use frames, use this document type declaration:

```
<!DOCTYPE HTML PUBLIC "-//W3C//DTD HTML 4.0 Frameset//EN" "http://
www.w3.org/TR/REC-html140/frameset.dtd">
```

Selecting New Document on the HTML Tools palette brings up a BBEdit build window. The author's first choice is to select DTD <!DOCTYPE> — the four options are HTML 3.2, HTML 4.0 Transitional, HTML 4.0 Frameset, and HTML 4.0 Strict. User-defined templates saved in BBEdit's templates folder (and identified in Preferences) are also accessible from this build window. The use of well thought out templates can minimize Web site development time and ensure code consistency.

Alternative Text

One of the simplest things a Web author can do to make sites accessible is to add ALT (alternative text) tags to images and other objects. This attribute tells the browser what to display if it is not possible to display the image (voice output or Lynx browsers, for example) or if the site visitor has disabled auto-load images.

The ALT attribute is part of the syntax of the APPLET, AREA, and IMG tags. The IMG tag is the most commonly used of the three. ALT is also optional for INPUT in the HTML-strict DTD where APPLET is deprecated. When a Web author uses BBEdit's image window to build an IMG tag, BBEdit adds the ALT tag even if the author only specifies an image source.

```
<IMG SRC="tools.gif" ALT="">
```

This fill-in-the-blank tag provides a visual clue that the author should add the HTML 4.0–required ALT tag contents. Note that syntax ALT="" is arguably useful for IMG when the image is used strictly as a spacer or decoration. However, many people prefer a short description included.

By selecting the image source from the Recent URLs or Current Folder pop-up windows, or by selecting File to browse drives (versus typing the file name and path), BBEdit also automatically inserts the Height and Width measurements, speeding up browser page rendering.

The ALT attribute is also part of the syntax of the APPLET and AREA tags. The ALT box is included in the APPLET build window, and BBEdit documentation explains why inclusion of this attribute is important for end users. Again, however, in the HTML 4.0–strict DTD, ALT is required syntax for the IMAGE element. It is not required for APPLET, a deprecated element.

Because the AREA tag is often used to map images to URLs (client side) for site navigation, including alternative text is essential when building navigable and accessible Web sites. BBEdit's AREA build window acts just like

II

8

the IMG window. Even if the Web author ignores the ALT box in the window, BBEdit includes it in the final tag:

```
<AREA SHAPE="rect" HREF="Tools.gif" ALT="">
```

As with the IMG tag, this code provides a visual clue that the HTML 4.0–required ALT tag contents are missing.

Fonts

Accessible Web sites set font sizes set as relative, not absolute, numbers. Although this element is being phased out in favor of Cascading Style Sheets, for backward compatibility, authors continue to specify font face and size.

The BBEdit font build window includes three elements compliant with HTML 4.0: face, size, and color. The author can easily specify font size by adding either a + or − to a number. However, merely typing in a number, color, or face is not sufficient to have BBEdit create the full tag; the author must also select the check box alongside each element.

Frames

Since the introduction of Netscape Navigator 2.0, Web authors have been able to divide a browser's main window into independent window frames, with each displaying different content. Frames were incorporated as a standard with HTML 4.0, with caveats about the need to make framed sites accessible to nonframe browsers.

Despite the W3C seal of approval, the use of frames remains a contentious issue and some graphical browsers (Opera for Windows and Microsoft Internet Explorer for the Macintosh, for example) enable the end user to turn off frames. Consequently, true user-friendly Web sites provide an explicit link to no-frames content. However, accessible sites must also include this link in the NOFRAMES portion of the FRAMESET.

Note

The WhatIs.Com Web site (http://www.whatis.com/) demonstrates an effective use of frames and provides explicit access to a no-frames version of the site.

What happens when a novice Web developer uses BBEdit to create a FRAMESET? The HTML Tools palette includes a FRAMES button; selecting it gives the author the option of selecting FRAMESET, FRAME, and NOFRAMES — the three tags required to build a FRAMESET. The initial selection enables the author to enumerate either the number of rows or columns. Next, the author specifies the source of the frame content. Finally, the author selects NOFRAMES, which provides an empty template.

```
<FRAMESET COLS="2">
<FRAME SRC="navigation.html" FRAMEBORDER="1">
<FRAME SRC="main.html" FRAMEBORDER="1">
<NOFRAMES>
</NOFRAMES>
</FRAMESET>
```

Thus, BBEdit does not require the Web author to include NOFRAMES content, nor is there an automatic signal (such as a blank ALT tag) to provide a visual clue that this information is needed. On the other hand, BBEdit documentation clearly states that this information should be included.

Only the three HTML 4.0 frame tags are included in the HTML Tools. The inline frame element is not included as a default choice, but it is included in the Tag Maker database.

Syntax Checker

With version 5.0, BBEdit's syntax checker uses tables derived from DTDs and checks against HTML 3.2 and HTML 4.0 Transitional, Frameset, and Final. The syntax checker uses the specification defined in the <!DOC­TYPE> SGML prolog at the top of the HTML document. If no <!DOC­TYPE> is specified, the default setting is HTML 4.0 Transitional. However, BBEdit issues a warning in this case, prompting the author to specify the <!DOCTYPE>.

The Web author can quickly select an element for editing or see if nested elements are formed correctly using the BALANCE TAGS feature. BBEdit sounds a system alert beep when it cannot find a matching set of tags around the selected text.

Other Features

Don't like the standard keyboard shortcuts? With BBEdit 5.0, you can assign a new keyboard shortcut to every menu command. Under the Edit

II

8

menu, use the Set Menu Keys command to assign a keyboard shortcut to any menu item, whether or not there is a preassigned shortcut.

For example, the default keyboard shortcut for images is Ctrl+cmd+I. However, it's a simple matter to change this to a two-key shortcut, such as, cmd+I. Because this combination is preassigned to debug (and BBEdit provides that warning), that setting must be overwritten. This is not an issue if the editor is being used strictly for HTML development. In this instance, debug has no assigned keyboard shortcut, and the author must create a new one.

CYBERSTUDIO and PAGEMILL CLEANER (accessible from the Markup Menu> Miscellaneous) remove font tags that have no semantic value (that is, font tags that surround white space) as well as the PageMill-specific NATURALSIZE FLAG attribute appended to IMG tags.

The HTML ENTITIES floating palette (accessible from the Windows menu) provides visual, one-click access to extended ASCII characters that are valid based on the current <!DOCTYPE>.

The Tag Maker (accessible from the Markup menu or cmd+M) is a new, context-sensitive HTML helper. Selecting Tag Maker launches a pop-up window that lists the valid HTML in the context of the insertion point (what attributes are valid for this tag) or the selection range (what tags can be nested inside this tag).

The WEB SAFE COLOR floating palette (accessible from the Windows menu) displays the 216 colors that display consistently in browsers running on all computers with monitors set at 256 colors. Use this palette to specify Web page background or font colors and improve the chances that each site visitor will have the same experience when viewing the site. Click the color and BBEdit inserts its RGB value into your document.

HomeSite

HomeSite (http://www.allaire.com/products/homesite/), a product of the Allaire Corporation, has a WYSIWYN (What You See Is What You Need) interface combined with standard HTML generation for Web authors using the Microsoft Windows platform. HomeSite is a highly customizable tool that provides support for JavaScript, ASP, Cold Fusion, Perl, and DHTML. Its user-defined color coding assists in visual debugging. The software was launched on the Web as shareware in July 1996, and currently boasts a user base of more than 200,000. It is the most highly acclaimed HTML text editor for the Windows platform. HomeSite's integrated development environment provides both tabbed-button and drop-down menu access to HTML tags (see Figure 8.5) and also provides easy access to graphics, scripts and other relevant HTML documents.

Note

The figures for HomeSite feature the HomeSite 4.5 evaluation copy. Note also that significant development is in progress to increase the accessibility of HomeSite in support of the WAI Authoring Tool Guidelines.

Figure 8.5   **The integrated development environment provides both tabbed-button and drop-down menu access to HTML tags**

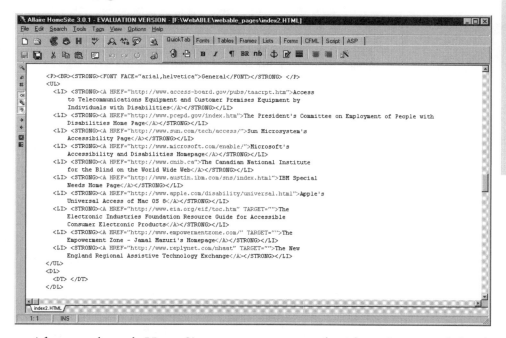

After you launch HomeSite, you are presented with an integrated development environment that includes a document editing window, a Windows Explorer–type view of files, and tabbed access to HTML tag sets. The default <!DOCTYPE> setting is HTML 4.0 Transitional (like BBEdit). HomeSite makes it easy to change the <!DOCTYPE> through a pop-up window. Contents of the window can be edited in the HomeSite Options>Settings>Default Templates window (see Figure 8.6).

Changing the <!DOCTYPE> is as simple as typing Ctrl+J and selecting HTML 2.0, HTML 3.2, or HTML 4.0 Transition from the pop-up window. HomeSite has created these keyboard shortcuts in advance. However, there is no shortcut for HTML 4.0 Frameset or HTML 4.0 Final, but they may be added to the list of menu items for this pop-up window.

Figure 8.6   HomeSite makes it easy to change the <!DOCTYPE> through a pop-up window

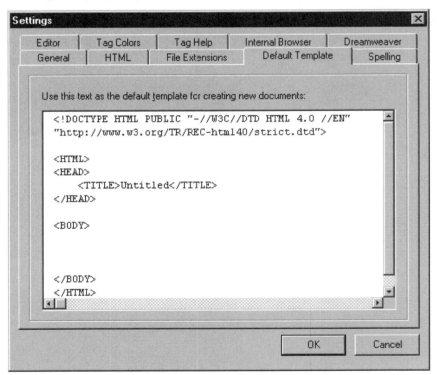

The most commonly used tags are included on the HomeSite tabbed toolbar, although tags are also accessible from the TAGS menu drop-down list.

Alternative Text

The simplest way to insert code for an image into a HomeSite document is to drag the image from the integrated Explorer window into the document window. Of course, you may need to edit the relative image path to match that on the Web server, but this drag-and-drop feature yields the image file name, height and width tags, and a blank ALT tag.

```
<IMG SRC="tools.gif" WIDTH="50" HEIGHT="50" ALT="" BORDER="0">
```

This fill-in-the-blank tag provides a visual clue that the author should add the HTML 4.0–required ALT tag contents.

Two other tags that require the use of alternative text are AREA and APPLET. Although the AREA tag is used to map images to URLs (client side) for site navigation, HomeSite does not have any information about the AREA tag in its help files; the coding for AREA is relegated to the Web author's image editing program. The button for the APPLET build window is on the Script tab, and it includes a text box for ALT information. There is also comprehensive information about the APPLET tag in HomeSite help files, including the ALT element.

Fonts

HomeSite provides two ways to specify font sizes: the font tag builder, or the larger or smaller font buttons on the Font tab. When using the font tag builder, HomeSite enables you to choose relative font sizes +1 through +4 and –1 through –3. These are the first to appear in the selection box, which prompts you to use relative sizes. Other elements included in the tag builder window are font face (only one choice allowed, and all system fonts are included), font color, and style sheet information.

The +/– Font buttons place only the size portion of the FONT set:

```
<FONT SIZE="+1"> </FONT>
```

Authors can complete the remainder of the tag by using the Tag Insight prompt or by right-clicking inside the tag element, which brings up the EDIT TAG window.

II

8

Frames

Like BBEdit's, HomeSite's designers expect the Web author to understand frameset fundamentals. The elements used to create a framed site are individual buttons on the tabbed toolbar. However, there is no dialog box for NOFRAMES (there is for the other three frame tags).

What happens when a novice Web developer uses the HomeSite frame wizard to create a FRAMESET? The visual interface enables the author to pick the number of columns and rows, and then provides text fields for frame NAME and SRC. There is no entry for NOFRAMES. Leaving all other options at default values yields the following code, which flunks HTML 4.0 validation on two points: no NOFRAMES content and "no" is not a valid attribute for FRAMEBORDER:

```
<FRAMESET COLS="2">

<FRAME NAME="navigation" SRC="Tools.gif" MARGINWIDTH="10" MARGIN-
HEIGHT="10" SCROLLING="AUTO" FRAMEBORDER="no">

<FRAME NAME="main" SRC="main.html" MARGINWIDTH="10" MARGINHEIGHT="10"
SCROLLING="AUTO" FRAMEBORDER="no">

</FRAMESET>
```

Thus, HomeSite doesn't even provide a fill-in-the-blank NOFRAMES placeholder, so there are no visual clues for the designer. And even though running the validator should yield an error message, it doesn't, whether <!DOCTYPE> is specified as HTML 4.0 Transitional, Frameset, or Strict.

Syntax Checker

HomeSite 4.0 has added validation options for HTML 4.0, but it also includes validation for Microsoft Internet Explorer and Netscape Navigator (the default is off for these proprietary extensions — see the Options>Settings>Validation window). The Web author can specify the type of information to return (including nesting errors), check for high ASCII characters, and check for quotes in text.

Other Features

Don't like the button choices on the tabbed toolbar? Just right-click and select Customize to add or delete buttons. Useful, but absent, buttons include Comment and Bold. Note, however, that for accessibility purposes, Bold and Italic are presentational elements. Be sure to use STRONG and EM instead. You are also able to edit keyboard shortcuts by selecting Options>Customize>Keyboard Shortcuts.

In addition, HomeSite provides detailed reference documentation not only for using the software, but also for HTML. For example, the Frames section details some reasons for using the NOFRAMES tag.

Figure 8.7   HomeSite provides easy access to HTML reference information when a full installation is performed

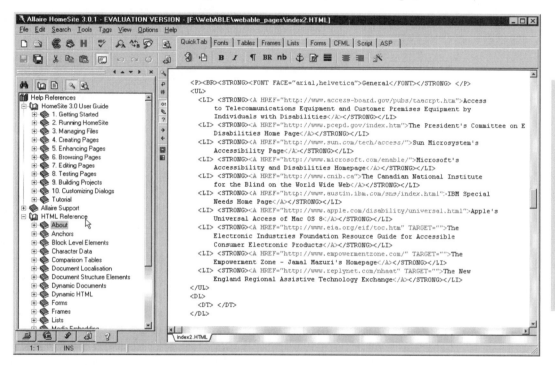

The SPECIAL CHARACTERS window (accessible from the View menu) provides visual, one-click access to extended ASCII characters.

TAG VALIDATION, TAG COMPLETION, and TAG INSIGHT (accessible from the T-bar separating the editing window from the Explorer interface) provide context-sensitive HTML help.

The WEB SAFE COLOR palette (accessible from the Button toolbar) displays the 216 colors that display consistently in browsers running on computers with monitors set at 256 colors. Use this palette to specify Web page background or font colors, and improve the chances that site visitors will have the same experience when viewing the site. Click the color and Home-Site inserts its RGB value into your document.

Update

Prior to releasing this book for publication, Sausage Software (HotDog Professional V6.0) and Chami.Com (HTML-Kit V1.0) released versions of their authoring tools that include accessibility features. Refer to their web sites for additional information. WebABLE also maintains a directory of authoring tools and their accessibility status.

Visual Editors

The term WYSIWYG (What You See Is What You Get) was born when Aldus Corporation and Apple teamed up with PageMaker, the first computing tool that enabled an author to manipulate text and graphics on a personal computer and have the screen look exactly like the finished page would. Graphical HTML editors use a similar interface, building the HTML in the background while the Web author concentrates on design and content.

Visual HTML editors insulate the Web author from the HTML code by providing a graphical development interface. As a result, the tools tend to favor appearance (physical formatting) over document structure. In contrast, HTML is grounded primarily in specifying document structure.

The current generation of graphical HTML editors (erroneously referred to as WYSIWYG editors) have made strides toward producing HTML that is compliant with current standards. However, there remains room for improvement, particularly in the area of prompting the Web author to incorporate accessibility features. Moreover, all graphical HTML editors tend to produce bloated code, often illustrated by redundant FONT tags that remain after copy has been deleted.

SoftQuad Software's HoTMetaL PRO Web publishing and site management application has established the standard for implementing accessibility features. Beginning with Version 4.0, HoTMetaL PRO included an accessibility checker, automatic accessibility prompting, and a Visual Dynamic Keyboard (VDK) to aid users with physical disabilities.

Two other popular and cross-platform graphical editors are Macromedia's Dreamweaver and Microsoft's FrontPage. Dreamweaver has an HTML viewer but also bundles either BBEdit or HomeSite as its text editor of choice. FrontPage also contains an HTML viewer.

Neither product produces HTML 4.0–compliant code out of the box. FrontPage generates extraneous code. Both require the author to understand the need for adding accessible code. However, Dreamweaver incorporates the text editors referenced in the previous section, which provide both syntax check and compliance tips.

HoTMetaL PRO

HoTMetaL PRO (http://www.sq.com) is a Web publishing and site management application produced by newly renamed SoftQuad Software (formerly SoftQuad). HoTMetaL PRO is a model Web publishing tool, and it is the only available Web publishing tool that includes embedded accessibility features. These features include:

- Accessibility prompting
- Accessibility checking
- Onscreen keyboard for the physically challenged
- Pop-up warnings

All of these features are optional. They are part of SoftQuad's AdaptAble Technologies, a collaborative effort between SoftQuad and ATRC at the University of Toronto. The AdaptAble Technologies are available in HoTMetaL PRO Versions 4.0 and 5.0. HoTMetaL PRO is currently available for Microsoft Windows 95/98 and NT.

HoTMetaL PRO is accessible to people who are blind and use Windows screen readers.

Yuri Rubinsky and SoftQuad

Frankly, when it comes to discussing SoftQuad's support for accessible Web publishing, I tend to lose objectivity on several levels. I'm not at all surprised that they were the first company to produce an accessible Web publishing tool.

SoftQuad's Cofounder and President, Yuri Rubinsky, who passed away in 1996, was chiefly responsible for launching the Web accessibility effort with the World Wide Web Consortium. Yuri made sure I went to all the WWW conferences to meet W3C staff and industry movers and shakers. And it was Yuri who encouraged me to start conducting the first workshops on Web accessibility at these conferences. We would spend long hours late at night trying to work on the accessibility implementations for HTML (the first work being available in HTML 2.0), using the old ICADD (International Committee for Accessible Document Design) SGML mechanisms.

On the day that he died, Yuri and I were talking via e-mail about the accessibility of HoTMetaL and Panorama, two of SoftQuad's products. He had to be sure that his products were accessible to all people. He lived and breathed accessibility.

Yuri Rubinsky and SoftQuad

Yuri Rubinsky

Not long after he passed away, several friends of Yuri, his wife, and his parents established a foundation in his name, the Yuri Rubinsky Insight Foundation (YRIF). I had the privilege of serving as the first (and only) director of the YRIF. Yuri's work was so instrumental to Web accessibility that those of us who comprised the original Web Accessibility Initiative (WAI) project team seriously considered naming the program office after him.

Unfortunately, the YRIF was forced to close shop in 1999. But, Yuri's memory lives on. We will never forget him.

Accessibility Prompting

You can set the level of accessibility prompting within HoTMetal PRO with the Automatic Accessibility Prompting menu item in the Options dialog box. Select one of the three radio buttons, whose level of accessibility is specified as follows.

Off No support for accessibility prompting. However, all the accessibility dialog boxes are available through HoTMetaL menu items.

On Elements requiring alternative text and/or descriptive text are prompted for, including images and image maps.

Strict Accessibility prompting is supported, as well as the pop-up warning system. Pop-up warnings are displayed in conjunction with the Insert Element dialog box for the following elements:

- TABLE
- BGSOUND
- FRAMESET
- MARQUEE
- BLINK

Figure 8.8 HoTMetaL PRO enables you to specify levels of accessibility settings in the Options menu

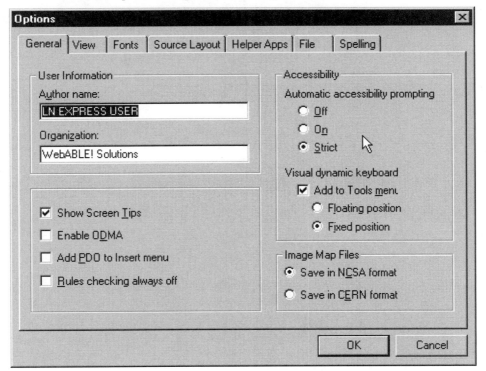

Entering Alternative or Descriptive Text

When you decide to insert an image on your web page, HoTMetaL PRO displays the Image Properties property sheet. Image Properties includes options for creating alternative text and/or descriptive text. If the image already exists and you simply want to add the alternative/descriptive text, double-click your mouse on the image and the Image Properties sheet appears.

To create alternative text, enter a short, meaningful description in the Alternate Text text entry box.

Figure 8.9   Use HoTMetaL's Image Properties dialog box to specify alternative text

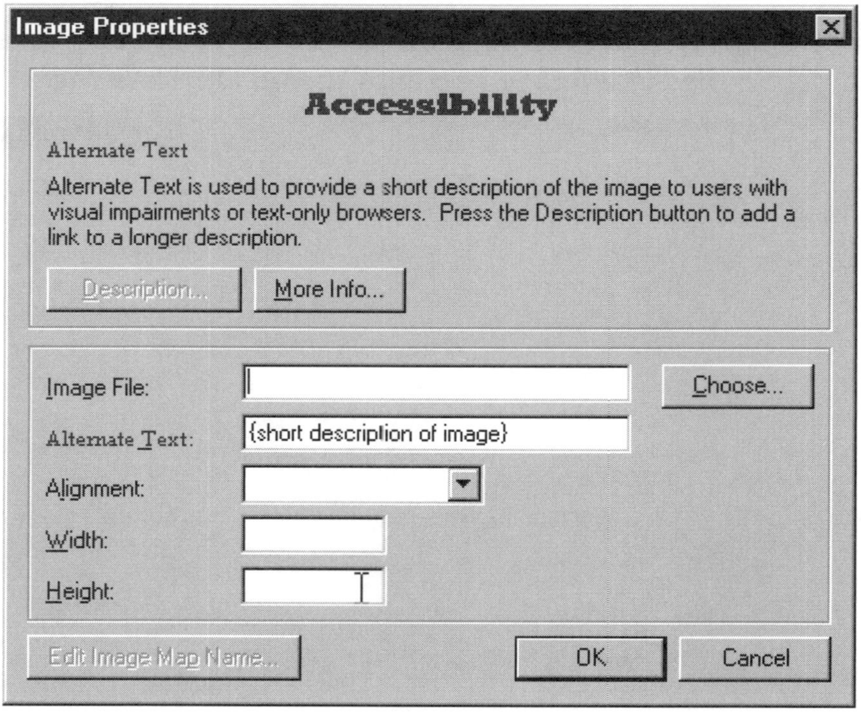

Imagine walking through a historical museum that includes an interesting display of dinosaurs. When you look for a description of the dinosaur you are viewing, all you see is a label with the word *Raptor* on it. Functionally this is correct, but likely you would want to know more about the Raptor. Where did it live? What did it eat? How big was it? This kind of information that gives you a better description of the Raptor and its lifestyle.

On the Web, this is what descriptive text provides. When you include an image or other graphical entity, providing descriptive text that describes the image's function and purpose helps an individual who is not able to view the image understand more about what its purpose is and why the image is there. In HTML 4.0, the LONGDESC attribute is used to provide a separate link to a file that includes the descriptive text. Unfortunately, no browser supports the LONGDESC implementation at this time. (See Chapter 5 for more information on how to implement LONGDESC.)

However, HoTMetaL PRO includes an excellent feature called Description that assimilates the function of the LONGDESC attribute and is supported by all graphical browsers.

To create descriptive text for an image, perform the following steps:

1. Choose Image from the Insert menu. (Double click an existing image.) The Image Properties property sheet displays.
2. Click the Description push button. The Accessibility Description dialog box appears.
3. Type the description of the image into the Image description text entry box.
4. Click OK.

II

8

Figure 8.10   Use the Accessibility Description dialog box to enter descriptive text

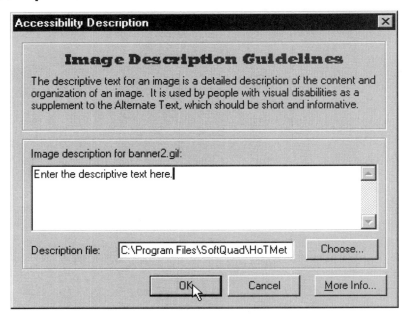

5. At this point, HoTMetaL PRO creates a new HTML file that includes your text description. It also creates a hyperlink to that file and associates it with the image. In the browser, the hyperlink is displayed as a lowercase *d* and appears next to the image on the Web page.

Figure 8.11 Descriptive text appears as a hyperlinked lowercase d in the browser window

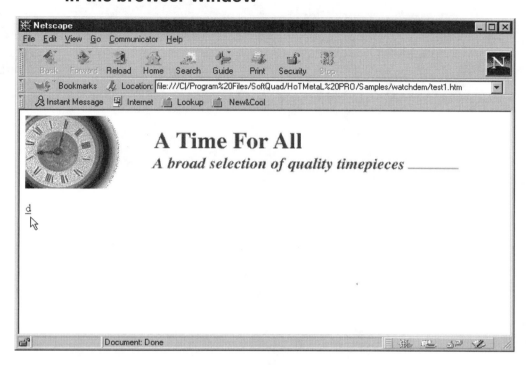

Accessibility Validation

To validate your Web pages for accessibility, choose Check Accessibility from the Tools menu. HoTMetaL PRO reviews your Web page and issues a summary statement of the errors. When the summary box is displayed, click the OK button and HoTMetaL PRO then reviews each error, giving you the opportunity to correct the error. To correct an error, click the Apply button.

Figure 8.12 Use the Check Accessibility to validate your Web pages

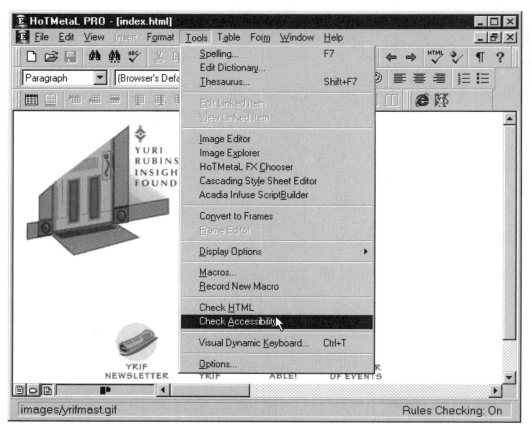

Note

HoTMetaL PRO does not specifically use the WAI Content Guidelines to check the accessibility of your Web pages. However, much of what it does check for is specified in the guidelines.

Visual Dynamic Keyboard (VDK)

To my knowledge, no mainstream application (other than some operating systems) includes a specific accessibility entity like SoftQuad's Visual Dynamic Keyboard or VDK. The VDK is an onscreen keyboard that enables users to enter text, select commands, activate dialog boxes, and otherwise fully operate the application using standard keyboard commands. The VDK is particularly useful for people with physical disabilities or those who otherwise cannot use a standard input device.

Figure 8.13 The Visual Dynamic Keyboard

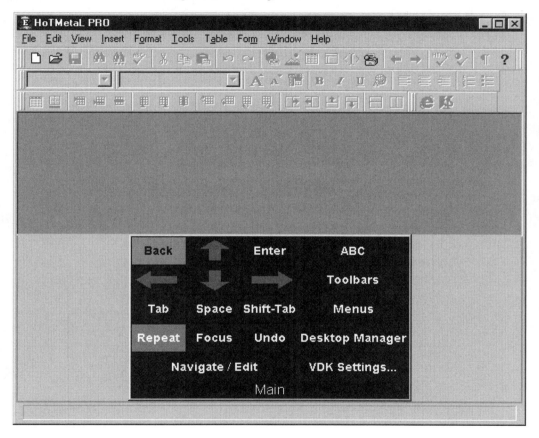

Dreamweaver

Macromedia's Dreamweaver 2.0 (http://www.macromedia.com/) incorporates Roundtrip HTML technology to import HTML documents without reformatting the existing code. It is a powerful authoring package that supports Cascading Style Sheets, JavaScript, and DHTML. Web authors need basic knowledge of accessibility to generate fully compliant code, but doing so does not present insurmountable obstacles. Dreamweaver is available on both the Microsoft Windows and Mac OS platforms.

After you launch Dreamweaver, you see a floating Object palette to the left of the visual document window. Another floating palette to the right launches windows and tools, such as the integrated HTML editor. A quick look at the basic code shows no <!DOCTYPE> designation. To add this without having to type it all from scratch, you have to select HTML from the launch palette, and then select BBEdit or HomeSite to get into the text editor that enables semiautomatic placement of this information.

Unlike most visual HTML editors, Dreamweaver includes an HTML inspector window that provides a real-time view of the raw code. Text in this window can be edited, but extensive edits and validation should be performed with BBEdit or HomeSite.

Alternative Text

To insert an image into a Dreamweaver document, first place the cursor at the point on the page where you want the image to appear. Then either select Insert>Image from the menu bar, click the Image button in the Object palette, or use the keyboard shortcut. The image is immediately visible in the document window.

Now look at the raw code:

```
<img src="tools.gif" width="50' height="50">
```

There is no ALT tag placeholder, nor is there a prompt to educate you as to its lapse. To add the ALT tag, you must know to now edit Image Properties (F3) in a small window at the foot of the document window. The image properties palette includes vspace and hspace, as well as low source, target, and link boxes. There is a drop-down menu for alignment that includes nine choices — most of which are nonstandard HTML.

Figure 8.14   Dreamweaver's menus often offer a prominent text box to enter ALT text

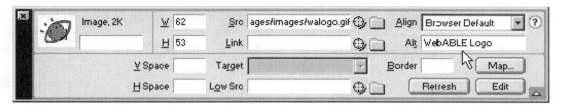

The image map editor, on the other hand, has a prominent text box for the ALT tag alongside text boxes for link and target. The APPLET properties editor also has a text box for the ALT tag.

Fonts

To edit content, either type (or paste) copy into the document window and then modify it from the Text menu or from the Properties palette. Dreamweaver groups font faces into five categories that accommodate preinstalled fonts available on Windows and Macintosh systems. It's easy to create relative font sizes by selecting Size Increase or Size Decrease off the Text menu but even easier using the Properties palette. However, the resultant code could be erroneous, as Dreamweaver enables you to set font size at +/–7. With a base font default of size=3, –2, and +4 are the legal parameters of the element.

II

8

Figure 8.15 Use Dreamweaver's Text menu to easily create relative fonts

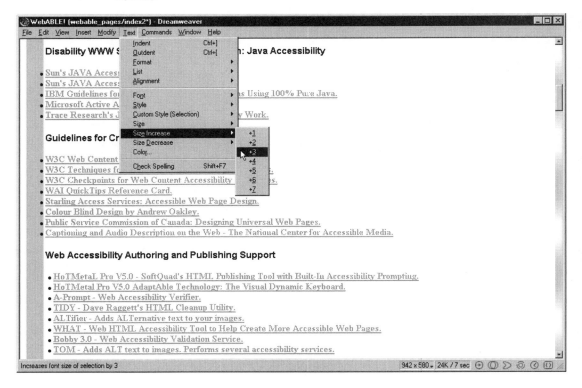

Frames

To create a framed site in Dreamweaver, choose Modify>Frameset from the main menu bar. This tool enables you to create two frames at a time. Alternatively, you can choose View>Frame Borders. This option enables you to drag a frame border and split the document window either vertically or horizontally.

Set frame properties using the Properties palette below the document window. Choices include frame name, frame source, scroll, resize, borders, border color, margin width, and margin height.

The `NOFRAMES` default is the same as BBEdit's — that is, there is a place-holder but no content. You can create NO `FRAMES` content in Dreamweaver by using the Modify>Frameset>Edit No Frames Content selection.

```
<noframes><body bgcolor="#FFFFFF">
</body></noframes>
```

Syntax Checker

Dreamweaver relies on the syntax checker that ships with BBEdit and HomeSite. Using the syntax checker integrated with BBEdit, the Dream-weaver document that has no `ALT` tag and no `DOCTYPE` produces requisite error messages. HomeSite did not reject the document that failed to include `NOFRAMES`, and it failed to highlight the error of including `FRAMESET` tags inside `BODY` tags.

Other Features

Dreamweaver provides one-click access to the Web-safe color palette. More-over, if you use the eyedropper tool to pick up a color off the desktop, Dreamweaver converts it to the nearest browser-safe color.

The program is preconfigured with a JavaScript library (including roll-over effects, form validation, and alert messages) that includes browser-detect code. It also automates creation of DHTML and Cascading Style Sheets. The software also imports and exports template content as XML.

FrontPage

Microsoft produces both a Macintosh and Windows version of FrontPage. Currently, the Windows version is FrontPage 98 and the Macintosh version is FrontPage 1.0. The tool is composed of two programs, FrontPage Explorer and FrontPage Editor. Explorer is used to create and maintain sites, and the Editor is used to create individual pages that make up a site. This review focuses on the Editor.

After launching FrontPage, before getting access to the Editor you must create a new FrontPage Web (or open an existing Web). Dialog boxes walk you through the process. It's now possible to launch the Editor by dou-ble-clicking a Web page icon. The feel is very much like a word processor, which can communicate a false sense of control over the appearance of the output.

Figure 8.16 Click on the HTML tab to view code generated by FrontPage

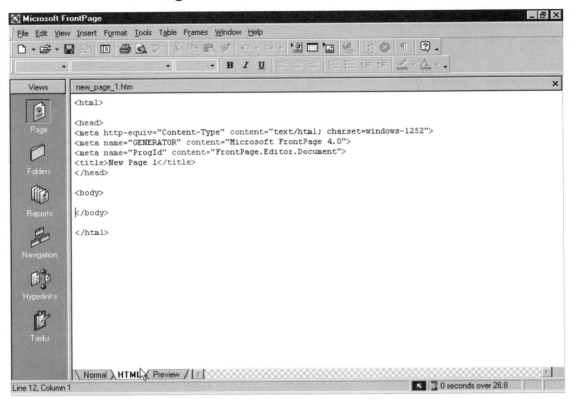

Using the Normal template to create a new Web and new pages, FrontPage inserts the following non–HTML 4.0 <!DOCTYPE> information in the template:

```
<!DOCTYPE HTML PUBLIC "-//IETF//DTD HTML//EN">
```

Alternative Text

To add an image to the Web page, click Insert>Image from the main menu bar or click the Image button on the toolbar. The Image dialog box displays folders and images. Select the image and click OK to embed the image in the

HTML. But what does the HTML look like? Is there an ALT tag place-holder? No.

```
<img src="Black_Squiggle.gif" width="25" height="15">
```

There isn't even a placeholder to alert you that a necessary element is missing. Adding the ALT text for IMG requires opening the Image Properties dialog box (right-click the Image with Windows, select Edit>Image Properties with the Macintosh), and then typing the text in the Alternate Representations area. The dialog box includes the Netscape-specific LOW-RES attribute.

The Insert Java Applet dialog box includes a message for browsers without Java support text field. Like HomeSite, FrontPage does not provide information about adding ALT tags to images used for clickable image maps.

Fonts

As with Dreamweaver, inserting text in the FrontPage Editor begins with clicking where you want the text to start, and then typing (or pasting) text. Changing font size can be accomplished from the Format>Font menu or by clicking the increase or decrease text size buttons on the toolbar. Again, there is a false sense of author control, as the (absolute) sizes are aligned with point sizes — that is, font size=1 is shown as 8-point type, which may or may not be the case, depending on the site visitor's default settings. Also, the menu box does not allow for relative font sizes.

In addition, all system fonts are accessible in the Font Face selection box. Thus, it is very easy for the uninitiated Web author to specify fonts that do not appear on the site visitor's computer, which can have a negative impact on readability and usability.

Frames

To create a frameset with FrontPage 1.0 for the Macintosh, select New Frames Page from the Frame menu. This brings up a dialog box with templates. The visual creation process is similar to Dreamweaver. And what does the default frameset code look like?

```
<noframes>
    <body>
    <p><!--Webbot bot="PurpleText"
    preview="The frameset on this page can be edited with the FrontPage
Frames Wizard; use the Open or Open With option from the FrontPage
Explorer's edit menu. This page must be saved to a Web before you can
edit it with the Frames Wizard. Browsers that don't support frames
will display the contents of this page, without these instructions.
Use the Frames Wizard to specify an alternate page for browsers with-
out frames."
    s-viewable=" " --> </p>
    <p>This Web page uses frames, but your browser doesn't support
them.</p>
    </body>
    </noframes>
```

Note that the default NOFRAMES content does not provide useful informa-tion to the frames impaired; however, there is an alert (the Webbot com-ments) notifying you of this step. Unfortunately, it is a multistep and nonintuitive process. In any case, the resulting code relies on proprietary FrontPage extensions for execution and retains the annoying "your browser doesn't support frames" message:

```
<noframes>
    <body stylesrc="noframes.html">
    <!--Webbot bot="Include" tag="BODY"
    u-include="noframes.html" startspan --><p> </p>
    <p>This Web page uses frames, but your browser doesn't support
them.</p>
    <!--Webbot bot="Include" erdspan i-checksum="48227" ->
    </body>
    </noframes>
```

Syntax Checker

There is no integrated syntax checker in FrontPage.

Other Features

FrontPage provides one-click access to a variety of color palettes. Unfortunately, the browser-safe palette is not at the top, so you must know to scroll down to look for this cross-platform tool.

To add special characters, select Insert>Symbol off the menu bar, select the character you want to add, and then click Insert.

FrontPage also includes the nonstandard HTML Blink and Marquee features, which are annoying to many Web site visitors (the Stop button on the browser doesn't affect these), and which can impede site accessibility for people with disabilities.

Amaya

As mentioned earlier in Chapter 7, Amaya is the World Wide Web Consortium's (W3C) own Web browser that also doubles as an authoring tool within the same window. Amaya serves an important function in the web development arena because the W3C passes many of its new Web-based protocols through Amaya. For example, Amaya already provides support for CSS1 and MathML. Amaya is available for UNIX and Windows 95/NT.

Amaya includes a WYSIWYG (What You See Is What You Get) editing interface. Assuming you have write access to a particular Web page, you can first browse a Web page and then edit it within the same window simply by single-clicking within the editing window. However, the authoring interface is not accessible to windows screen readers.

To create alternative text for an image or image map using Amaya, click the Types menu and select the Image (IMG) menu item. The Open Document dialog box is displayed. Enter a brief, meaningful description of the image in the Alternate Text text entry box.

Amaya includes support for many new HTML 4.0 elements and attributes that increase the accessibility of a Web page. For example, users with visual disabilities who rely on speech output of a Web page often find lengthy forms difficult to navigate and understand. In HTML 4.0, the OPT-GROUP element can be used to logically group form choices. To do this in Amaya, simply click the Types menu, select Forms, and then select OPTGROUP from the list.

Converters

As corporations move their employee communications to intranets, the demand for easy conversion of basic (and legacy) documents to HTML grows. Proprietary document formats have been the bane of corporate existence since the dawn of the computer age — the open standards of the Web provide another chance to break from proprietary formats (rich text format [RTF] is another effort).

HTML converters for Microsoft Office products are built into Office 97 and Office 98 suites. There are also converters for Adobe PageMaker, QuarkXPress, and Corel WordPerfect, to name a few.

Currently, all converters are handicapped by the lack of structure in the original document. Consequently, the code must focus on presentation, or appearance. This focus is opposite of the philosophy of HTML, which specifies document structure. As a result, the HTML generated often contains not only syntax errors, but also is usually even more bloated than code created by a visual editor.

Before publishing for general Web viewing, code from converters should be checked with an HTML validator and then hand-coded (with a text editor or a text HTML editor) for clean up. *Do not reopen the document with the original converter.* More often than not, the good code becomes corrupted.

Another entry in this category could be IBM's Domino, which provides Web publishing of Lotus Notes information. For companies with extensive Notes databases, Domino can provide a cost-effective way to move the information onto the Web (internal or external). Still, caveats remain — the first generation of Domino, for example, did not use the browser-safe color palette and allowed little direct editing (corrections) of code for HTML 3.2 compliance.

As Cascading Style Sheets become more prevalent, converters may be able to read styles from native documents and convert this presentation information to CSS.

Guidelines for Creating Accessible Authoring Tools

Creating accessible Web sites would be a much easier task if authoring tools and publishing suites supported accessibility within the application environment. Implementing wizards, validation services, strict HTML 4.0 support, and a comprehensive context-sensitive help system would go a long way toward ensuring access to people with disabilities.

In February 2000, the WAI Authoring Tool Accessibility Guidelines 1.0 became an official World Wide Web Consortium (W3C) recommendation. You can access the guidelines at `http://www.w3.org/TR/WAI-AUTOOLS/`.

The goal of the guidelines is to help developers create authoring tools that produce accessible web content. Additionally, the guidelines recommend methods for making the publishing tools themselves accessible to people with disabilities. In total, there are 7 primary guidelines and each guideline includes a list of checkpoints. The 7 guidelines that authoring tool developers are encouraged to implement in the publishing tools are:

- Support accessible authoring practices
- Generate standard markup
- Support the creation of accessible content
- Provide ways of checking and correcting inaccessible content
- Integrate accessibility solutions into the overall "look and feel"
- Promote accessibility in help and documentation
- Ensure that the authoring tool is accessible to authors with disabilities

Summary

The World Wide Web is built on a foundation of flexible, user-directed display. The ideal Web development tool focuses on standard tags, provides warnings when the Web author is about to embark on nonstandard code, and is customizable. Currently, HTML text editors do the best job of meeting these goals.

Web tools cognizant of these concepts help authors build Web sites that support the concept of universal and accessible Web design, which in turn supports the growing variation in Web display and access. For example:

- Hands-free, voice-activated browsing devices such as Web phones
- Variations in display, from WebTV to hand-held computers (such as the PalmPilot) to Web kiosks
- The large number of end users with slow Web connections, particularly among non-U.S. citizens
- Web users who prefer text-only browsing to avoid blinking animated ads or to facilitate download time
- A rapidly growing aging population, with the accompanying decrease in visual, hearing, and motor skills
- The relatively high Web presence of people with sensory and motor disabilities

References

A-Prompt Tool Kit Web site. `http://aprompt.snow.utoronto.ca/`.

Allarie's HomeSite Web site. `http://www.alliare.com/prdocuts/homesite/`.

BareBones Software Web site. `http://www.bbedit.com/`.

Macromedia Web site. `http://www.macromedia.com/`.

Microsoft Corporation Web site. `http://www.microsoft.com/`.

SoftQuad Software Web site. `http://www.softquad.com/`.

World Wide Web Consortium. "HTML 4.0 Specification." Available from `http://www.w3.org/TR/REC-html40/`.

World Wide Web Consortium. "Web Authoring Tool Accessibility Guidelines." Available from `http://www.w3c.org/TR/WD-WAI-AUTOOLS/`.

World Wide Web Consortium. "Web Content Accessibility Guidelines." Available from `http://www.w3.org/TR/WAI-WEBCONTENT/` `http://www.w3c.org/TR/WD-WAI-PAGEAUTH/`.

9

Chapter 9

Programming for Web Accessibility

In This Chapter

- Java accessibility programs
- The Java Speech API
- Microsoft Active Accessibility

This chapter is intended to provide you with an overview of programming features and techniques that can enhance the accessibility of user interfaces in general, and those found in Java applets and applications specifically.

Web Programming and Accessibility

The Web is well beyond the stage of static, change now and then web sites. The advent of content that is coded in XML, stored and driven by massive computer databases, and served up as applications is here. This is also means that many software vendors are beginning to move from the production and distribution of off-the-shelf software products to web-served applications that are accessible through every possible communications channel – wireless, cable, DSL.

Not only is this the next major paradigm shift in the software industry (next to the development of voice enhanced applications), it clearly presents the next major hurdle for web and Internet accessibility. Presently, most assistive software cannot interpret or otherwise render web content that is scripted or programmed. Unfortunately, this is a bi-product of niche market product development. Today, specialized web clients like HomePage Reader and pwWebSpeak ignore web content that includes scripts and most programming language constructs. Subsequently, the user is not likely to know that this content or application exists on your site.

Companies like E*Dapta (`http://www.edapta.com`) are trying to bridge the gap by providing a middleware transformation service. In other words, the software transforms the web content and/or application to the user based on previously established preference settings defined by the user. Internet and application service providers will likely license the technology providing thus providing a reasonable level of accessibility for their subscribers.

However, as I've stated from the beginning of this book, the real solution is to build applications and web content (regardless of how it's stored or rendered) that is inherently accessible. While this may seem tedious, it's not impossible and certainly not nearly as difficult now that Sun and Microsoft have developed software architectures and API's that support accessibility. Both of these are described in the sections that follow.

Java Accessibility Programs

Sun Microsystems is currently focusing on incorporating accessibility into several main areas of the Java platform: the Java Accessibility API, Java Accessibility Utilities, and the Java Accessibility Bridge to Native Code.

The Java Accessibility API Technical Overview

The Java Accessibility API is a part of the Java Foundation Classes and is designed to give assistive technologies access to information in user interface objects. Java applets and applications may be written to support the Java Accessibility API. If one does, screen readers and other accessibility aids will likely be compatible with the application.

The API contains eight Java programming language interfaces and six Java programming language classes.

Programming Language Interfaces

Following is a list of the accessibility programming language interfaces:

Interface Accessible   All components that support the Java Accessibility API are required to implement this interface.

Interface Accessible Action   Any object that can be manipulated with actions should support this interface. It enables the assistive technology to determine what actions can be performed by an object, and to direct those actions when invoked by the user.

Interface AccessibleComponent   Provides the means for assistive technologies to graphically represent objects.

Interface AccessibleSelection   Tracks the current selection set of children of an object. It also enables modification of that selection.

Interface AccessibleText   Implemented when an object contains rich, editable text. This isn't simply any text displayed, but text that may be modified by the user. It can operate on text appearing at a given pixel coordinate, letters, words, and phrases, as well as the attributes of individual characters such as font selection, italics, and so on.

Interface AccessibleHyperText   Any object that displays hypertext or enables for the activation of hyperlinks must use this interface. HTML element attributes and the element contents are exposed to the interface for interpretation by the assistive technologies.

Interface AccessibleValue   Any object that supports numerical values, including those that control pixel-level position by coordinates.

Programming Language Classes

Following is a list of the accessibility programming language classes:

Class AccessibleContext   This class returns the basic minimum information about an accessible object, including its name, current state, description, and relationships as a parent or child to other objects. Methods native

II

9

to this class can also enable technologies in obtaining more granular information about a given component.

Class AccessibleRole   The role of an accessible object in the user interface is trapped using this class. A list of predefined roles exists, such as check boxes or radio buttons, but the class does enable extension with programmer-supplied roles.

Class AccessibleState   The state of an object may include "selected" for radio buttons, "checked" for check boxes, "focus" for the button that has the current input focus, and more. As with AccessibleRole, a predefined list of states exists, with the option to customize.

Class AccessibleStateSet   The complete set of states present in one object. A radio button may have the focus, yet be unchecked. The AccessibleStateSet contains information on each possible state that an object may have.

Class AccessibleBundle   The super-class of AccessibleRole and AccessibleState.

Class AccessibleResourceBundle   Not normally used by programmers.

Java Accessibility Utilities

Java Accessibility Utilities are a collection of utility classes for Java programming that enable assistive technologies to provide access to graphical user interface (GUI) toolkits that implement the Java Accessibility API. The current release of the utilities package supports the Java Development Kit (JDK) Version 1.2.

Note

At the time of this writing, the Java Utilities version for Java 2 SDK (was JDK) was released and could be downloaded from Sun's Web site at `http://java.sun.com/products/jfc/index.html#download-access`.

The Java Accessibility Bridge to Native Code

Individual operating systems such as Microsoft Windows, Mac OS, and UNIX already incorporate some accessibility features in their user environments. In order for the Java Accessibility features in Java applications to interact with those native abilities, some form of standardized communication must be developed. The Java Accessibility Bridge to Native Code spans that gap.

The bridge is a Java class that contains native methods. The operating system in use has its own Dynamic Link Library (DLL) or other programmatic collections of instructions that it uses to process accessibility features. That DLL communicates with the portion of the Java Accessibility Bridge class that is devoted to that operating system. That section of bridge class then interacts with the Java Virtual Machine and on out to the Java Accessibility API, the utility support, and the targeted user interface objects that should be manipulated in the native accessible manner. Figure 9.1 charts this interaction.

Figure 9.1   The flow of data using the Java Accessibility Bridge for Native Code

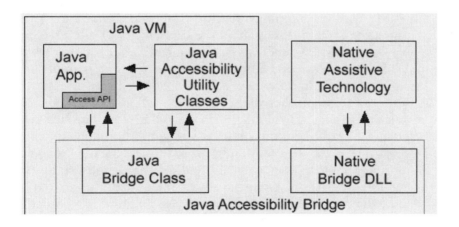

Guidelines for Creating Accessible Java Programs

Several Web sites are great resources to Java application developers and designers to assist in the development of accessible Java programs. A few key resources include:

- Sun's Java Swing Tutorial, "How to Support Assistive Technologies": `http://java.sun.com/docs/books/tutorial/uiswing/misc/index.html`
- IBM Guidelines for Writing Accessible Java Applications: `http://www-3.ibm.com/sns/accessjava.html#guidelines`
- Trace Research's Java Accessibility and Usability Site: `http://www.trace.wisc.edu/world/java/java.htm`

All three sites provide guidelines, resources, and several examples of Java application code that includes accessibility.

Java accessibility is catching on everywhere . . . even with Microsoft! The next section examines Microsoft Active Accessibility and support for Java.

Microsoft Active Accessibility and Java

The Microsoft Active Accessibility (MSAA) program was introduced in May of 1997.

MSAA is a technology developed by Microsoft that is integrated into the operating system. It enhances the capabilities of adaptive devices and software applications such as screen magnifiers or windows screen readers. While these devices and applications act as *accessibility aids* that enable enhanced or alternative presentations of the information found on the screen, Microsoft Active Accessibility can help an adaptive device distinguish between individual portions of the user interface such as toolbars, menus, alert boxes, and other visual screen components and their relationships to each other. As a result, programmers are able to extract more information about the user interface and make that information available to the accessibility aid. In turn, the user benefits from improved usability of the accessibility aid and often is able to use an interface, hardware, or software as productively and proficiently as a person who does not have a disability.

The Active Accessibility system is fully implemented in the Windows 98 operating system and in beta versions of Windows 2000. Users of these operating systems need no additional utilities in order for compatible accessibility aids to interact with the Active Accessibility systems. Users of earlier operating systems such as Windows 95 can install the Active Accessibility run-time module that may be distributed by the developer of individual aids. The recent release of Service Pack 4 for Windows NT provides that operating system with partial support for the technology.

Developers who wish to implement Active Accessibility in their software programs need the Active Accessibility Software Development Kit (SDK) and the run-time components that are packaged as the Redistribution Kit (RDK). The SDK includes documentation in several formats (including HTML and MS Help format), testing tools, sample source code, header files and libraries for programmers, and more. These kits, along with release notes and licensing agreements, can be obtained as a part of the Active Accessibility Client Developer Package. Developers must register for the Microsoft Active Accessibility Beta program to receive the developer package. You can find details at `http://www.microsoft.com/enable/msaa/msaabeta.htm`.

At the time of this writing, the SDK was being updated to reflect the changes that came with Microsoft Active Accessibility 1.2.

Technical Overview

MSAA is based on the Component Object Model (COM), a Microsoft technology that provides a standardized method of communication between software clients and the operating system.

MSAA takes a client-server approach to interaction. An application — for our purposes, a software program such as a Web browser — is known as the Microsoft Active Accessibility *server*. The *client* is the accessibility aid being used. The server (the Active Accessibility–compliant software program) gathers information about the contents of the windows (screens) that it controls. That information is passed to the client (the accessibility aid) as COM objects. Those objects, specifically referred to as accessibility objects, store information about the objects content and state (screen position,

II

9

focus, and so on.). The object also has native methods that may be programmatically called by the aid in order to perform specific functions.

Guidelines for Creating MSAA Applications

Software engineers and application programmers will benefit from the extensive set of guidelines provided by Microsoft for developing accessible applications. Refer to "The Microsoft Windows Guidelines for Accessible Software Design" which can be downloaded from `http://www.microsoft.com/enable/dev/guidelines/software.htm`. The guidelines are available in HTML and Microsoft Word file formats.

Briefly, Microsoft's guidelines focus on the following design principles:

Flexibility   Design flexible user interfaces that can be easily customized by users and accommodate their preferences.

Choice of input methods   In addition to simplified mouse operations, be sure to include full keyboard access to all features of the user interface.

Choice of output modalities   Adapt to the user's choice of output modalities, whether it's visual or sound.

Compatibility with accessibility aids   Ensure that all interface and programming code is accessible and compatible with accessibility aids, including screen readers, screen magnification, and voice recognition.

Consistency   Design consistently throughout the application, whether it's the interface or system behavior.

MSAA for Java

As Web-based applications become more pervasive through information applications and services, the need to develop highly usable and accessible interfaces will be crucial for all users. Developers already familiar with Microsoft Windows programming concepts, particularly Active Accessibility, will appreciate that MSAA now includes support for Java. MSAA for Java includes the Java-based interfaces required by Java developers to create applications that are accessible, seamless, and portable across any operating system platform that supports Java. In one full sweep you benefit from the cross-platform portability of Java's "write-once, run everywhere" technology and MSAA's architecture and interface standards that support accessibility aids.

Active Accessibility for Java is based on Microsoft's Application Foundation Classes (AFC), a set of Java class libraries providing user interface controls.

For additional information about creating Active Accessibility applications for Java, refer to Microsoft's Technologies for Java Web page at `http://www.microsoft.com/java/`. This site also includes Microsoft's software developers kit (SDK) for Java. At the time of this writing, this was SDK Version 3.2.

Java Speech API

Sun Microsystems collaborated with leading speech technology companies to develop the Java Speech API (Application Program Interface). The API provides Java developers with a software interface for integrating speech technology in their applets and applications. The Java Speech API was developed in collaboration with leading speech technology companies: Apple Computer, Inc., AT&T, Dragon Systems, Inc., IBM Corporation, Novell, Inc., Philips Speech Processing, and Texas Instruments Incorporated.

Two primary processes are addressed in the Java Speech API: *speech recognition* and *speech synthesis*. Speech recognition involves the computer application recognizing and converting the spoken word into commands and actions. Think *Star Trek* and the natural language interaction between crew and computer — ask a question or issue a command and the computer processes the request and respond appropriately.

Speech recognition systems that are available today include dictation software such as Dragon Systems' NaturallySpeaking or IBM's ViaVoice. With these systems, the user can speak with normal inflection at a comfortable pace, and the software converts the spoken word into text. The user is given a chance to review the conversion and correct any errors, thus helping the program to "learn" the user's speech patterns. In fact, portions of this chapter were written with such a product.

Another present-day application of speech recognition is the voice response systems used in voice mail and other telephony packages. Rather than having to press the 1 button on your touch-tone phone, such systems enable you to speak the word *one* and achieve the same result.

In the future we will likely move toward a complete independence from hardware based input devices, à la *Star Trek*. Improving human-to-computer interaction is a goal for all users. Hands-free computing has applications in a variety of environments — interacting with geo-navigation systems in automobiles, programming those pesky VCRs, and any number of household or industrial tasks that require computer interaction, but run the risk of contaminating an input device with grease or other undesirable substances. The utility of many of these applications has come to light based on the work performed by pioneers in the accessibility field.

Speech synthesis, the second major speech application, requires the computer to interpret text appearing in files or dynamically encountered in the application and then convert it to audio output. The screen reader browsers discussed in Chapter 7 are examples of applications that perform this text-to-speech process. These products can synthesize speech from text or data already present in the applet or application, read data delivered to it (such as in the Web browser example), or convert the user's input from text to sound.

Synthesized speech can improve many current systems that use prerecorded voice responses, such as the telephone account access services provided by many banks. Presently, the systems combine prerecorded phrases to produce the necessary response. You may have noticed how unnatural and choppy they sound. Speech synthesis technology will enable the system to formulate its response *after* it has gathered the data that must be returned to the user, providing for a much smoother and more understandable response.

Design Goals for the Java Speech API

Sun's stated goals for the Java Speech API, combined with other Java Media and Communications APIs, is to enable developers to incorporate advanced user interfaces into Java applications. The following design goals will make this possible:

- The Java Speech API will support speech synthesis, command-and-control recognizers and dictation systems.
- The Java Speech API will be simple and compact
- "Write Once, Run Anywhere" access to speech synthesis and speech recognition will be consistent across the major Java platforms and across products from different speech technology companies.
- Existing speech technology should be accessible through the Java Speech API using bridging software provided by Sun, licensees of Java, and others.
- The Java Speech API will complement other Java features, including the Java Media and Communication APIs.

Overview of Technical Issues

The Java Speech API covers three major areas of application development: resource management, speech recognition, and speech synthesis. Resource management refers to the interaction between the Java application and the audio systems of a computer, and acts as a controller for the speech recognition and synthesis routines.

Handling Speech Recognition

The core capabilities of the Java Speech API include the control of speech input, management of vocabulary, definition of grammars used for recognition, and the trapping of recognition results and other priority events. Two major grammar recognition systems are supported: *rule-based grammars* and *dictation grammars*.

When rule-based grammars are used, the speech recognition module receives hints from the application as to what type of input the user is expected to provide — in other words, the expected response phrase or words. If the application has been designed carefully —including cues to the user as to what responses should be, yet allowing enough flexibility that a negative response of "No" and "Nope" would both likely be interpreted correct — the speed at which recognition is performed can be managed as quickly as possible. Sun provides the following example of a simple command recognition structure:

```
RuleGrammar SimpleCommand;
    <COMMAND> = [<POLITE>] <ACTION> <OBJECT> (and <OBJECT>);
    <ACTION> = open | close | delete;
    <OBJECT> = the window | the file;
    <POLITE> = please;
```

Based on this simple rule, a user could make any of the following requests, and have the request recognized by the system:

- "Open the window."
- "Delete this file, please."
- "Please close the window."
- "Would you please delete that file?"

With significantly more complex rules (and correspondingly increased processing time or processing power required), the user will have more freedom to interact with the system at a natural language level.

Dictation grammars have historically required the user to interact with the system in *discrete* units of speech. Meaning, short pauses between words or short phrases are required for the system to be able to recognize what the user is saying. Unfortunately, this unnatural and stilted speech is difficult for the user to keep up for extended periods of time. *Continuous speech* dictation systems, which enable the user to speak at a more natural speed, have undergone recent improvements, and are featured in the Dragon Systems and IBM voice systems previously mentioned.

Compatibility with other APIs

The Java Speech API is being developed as a part of the Java Media and Communication family of APIs. This metaset is intended to provide a complete integration of multimedia functions within Java applications. APIs in the Java Media and Communication family that most related to the Java Speech process include the Java Media Framework, the Java Telephony API, and the Java Sound API.

Related Work

A core technology for speech synthesis is the annotation of text and input so that the application may appropriately form audio output. Sun has developed a unique markup language intended to facilitate this annotation, known as the Java Speech Markup Language (JSML). JSML and the Java Speech API handle different functions in speech processing. JSML describes the textual representation of input to a speech synthesizer. Sun makes it clear that JSML does not address the following issues:

- Mechanisms for providing marked-up text to a synthesizer
- Software control of the output of annotated text such as queuing, pause and resume, and variation of pitch and speaking rate
- Mechanisms for receiving notification of synthesis events including marker events requested in JSML text
- Error handling capabilities including incorrect markup
- Vocabulary management issues such as provision of pronunciations

The Java Speech API does address these listed issues, and complements the capabilities of JSML, providing a complete solution.

Summary

This chapter reviewed several major programs for incorporating accessibility features in Java applications. The Java Speech API also extends traditional programs and applications by enabling speech recognition and synthesis to be built into the user interface, thereby enhancing the experience for both disabled and able-bodied users.

References

comp.speech.newsgroup. "Comp.Speech Frequently Asked Questions." Available from http://www.speech.cs.cmu.edu/comp.speech/.

Sun Microsystems, Inc. "How to Support Assistive Technologies" (Java Swing Tutorial, 1999). Available from http://java.sun.com/docs/books/tutorial/uiswing/misc/index.html.

IBM Corporation. "IBM Guidelines for Writing Accessible Java Applications," 1998. Available from http://www.austin.ibm.com/sns/accessjava.html#guidelines.

IBM Corporation. "IBM ViaVoice." Available from http://www-4.ibm.com/software/speech/.

Jamie Jaworski. *Java 1.2 Unleashed — The Comprehensive Solution.* Sams, 1998.

Sun Microsystems, Inc. "Java Speech API." Available from http://java.sun.com/products/java-media/speech/index.html.

Sun Microsystems, Inc. "Java Speech API Programmer's Guide," 1997.

Sun Microsystems, Inc. "Java Speech Markup Language," 1997.

Microsoft Corporation. "Microsoft Active Accessibility for Developers, Writers and Designers." Available from http://www.microsoft.com/enable/dev/default.htm.

Microsoft Corporation. "Microsoft Java Accessibility." Available from http://www.microsoft.com/enable/msaa/default.htm.

Microsoft Corporation. "The Microsoft Windows Guidelines for Accessible Software Design." Available from http://www.microsoft.com/enable/dev/guidelines/software.htm.

II

9

Microsoft Corporation. "MSAA for Accessibility Aid Developers."

Sun Microsystems, Inc. "Summary of the Java Swing 1.1 API Specification." Available from `http://java.sun.com/products/jfc/swing-doc-api-1.1/javax/accessibility/package-summary.html`.

Part III

Development Resources

Chapter 10

Specialized Web Accessibility Software

In This Chapter

- Introduction to accessibility tools
- Operating system tools
- Screen readers
- Multimedia tools
- Other useful Web accessibility services

Because of the wide variance in human abilities, it's extremely difficult to design a human interface for a product such as the Web in a way that is completely accessible to all people, particularly in its first or second iteration. For example, while Web access to people with disabilities is a major issue, internationalization of the Web is an even larger one. Web content tends to be primarily available in English, though recent strides in translation services have made it easier to produce multilingual web sites.

Just as the international community is disparate, the community of people with disabilities is also disparate by nature; thus, it is almost impossible to make the World Wide Web completely accessible to every person with a disability. This is not an excuse to justify the inaccessibility of the Web — it's a fact of life. Within every community of disability there exists degrees of ability; not every person with a visual disability is totally blind, not all people with mobility disabilities are paralyzed, and certainly not all people who are hearing impaired are deaf. Therefore, it is incumbent upon a system administrator to become familiar with a handful of tools and utilities that provide additional assistance to a user with a disability.

In this chapter, I provide an overview of the available applications — mostly software — specifically designed to enhance and improve the Web experience for people with disabilities. There are several operating system–level features you can also use to simplify general computer use for a person with a disability.

Additionally, I cover screen readers, which represent a specific technology developed for people with visual disabilities. Screen readers are tools that use speech output technology to read back computer data to the people with visual disabilities. For the deaf, recent advancements in captioning applications have also been developed.

Most of the accessibility software and services I discuss were developed exclusively to aid people with disabilities. However, several mainstream applications — for example, authoring tools that implement captioning — are part of larger commercial product offerings.

Before I delve into the available tools, let's look at what accessibility software provides users with disabilities.

Introduction to Accessibility Tools

In Chapter 4, I briefly discussed access systems people with disabilities use to help them render or view Web content. This chapter discusses the tools and utilities — assistive and adaptive software — that individuals use as part of their everyday interaction with computers.

As a system or Web administrator, you should become familiar with these applications. It is just as important to ensure that the computing environment is accessible as it is to ensure that your Web site is accessible. In a corporate environment where intranets and collaborative applications are becoming increasingly popular productivity tools, understanding how an individual with a disability can use these tools is crucial. Remember, too, that many world governments mandate equality of access and reasonable accommodation for people with disabilities.

Types of Accessibility Tools

Screen readers, refreshable Braille displays, screen magnifiers, mouth sticks, bounce keys, eye-gaze systems, captioning software — these are a small fraction of the accessibility tools used by people with disabilities in their daily interactions with computers. Many of these tools are also used by people with disabilities while surfing the World Wide Web.

As mentioned in Chapter 4, Web accessibility tools include the following:

- Synthetic voice systems, digital audio, or Braille for the blind
- Screen magnification — large text fonts for those with diminished vision or dyslexia
- Descriptive text, captioning, and visual cues for the deaf or hearing impaired
- Specialized adaptations for the physically challenged and mobility impaired involving the use of a keyboard, mouse, or input device that requires a part of their body other than their hands and fingers to control browser or other user agents required to view Web content

Most of these tools are available as operating system preference settings and software applications. Some adaptations exist as hardware solutions — for example, a mouth stick is an assistive device inserted in a person's mouth (usually between the teeth) that extends down toward the keyboard. This device enables a user to type keyboard commands. (You can appreciate why it's important to ensure that your Web site is accessible to keyboards under these conditions.)

First, I'll discuss the operating system–level tools.

Operating System Tools

Operating system–level tools and utilities exist for the major operating environments including:

- Microsoft Windows 3.1/95/98/2000
- UNIX and X Windows (including Linux and Motif)
- Mac OS (7, 8.5, and 9.0)
- IBM OS/2 WARP

My experience has been that many system managers and administrators are not aware of some of the more commonly known and freely available accessibility features that operating systems provide. Implementing these features can help companies cut costs.

To illustrate, on one occasion I was asked to do an accessibility site review for a major university. To the credit of the university, they were making every effort to accommodate their disabled students and staff in spite of a limited budget. However, they were unaware that almost all their personal computing systems already contained several accessibility options. Likely the site administrators had turned the features off, never installed them, or uninstalled them. A few minor adjustments were all that was required to immediately enhance site accessibility.

Note

People often ask me which operating system is the best suited for people with disabilities. The answer is simple: the one that works best for you or the individual with a disability. Operating systems are like eggs — everyone has their preference. Some people like hardboiled; others scrambled. Still others like theirs over-easy. The same is true for operating systems. Some like Mac OS or UNIX, and others prefer Microsoft Windows. Even among Microsoft Windows users there are preferences between Windows 3.1, 95/98, and 2000.

No operating system is completely accessible, and I won't get caught arguing for one operating system over another. All of them contain measures of accessibility. Additionally, assistive technology software vendors create their products to be compatible with most of the major operating system platforms. The decision is yours.

The following sections provide a list of the various accessibility features and tools each operating system offers. I've tried to capture most of the features currently available on each platform. For a more in-depth description of operating system and application accessibility, wait for my next book!

The Software Gold Mine: Trace Research

Trace Research is *the* leading research facility for information technology involving people with disabilities, particularly with regard to software. On the Trace Web site (http://www.trace.wisc.edu/), there is an area specifically devoted to operating systems that you should bookmark in your Web browser (http://www.trace.wisc.edu/world/computer_access/). Current or obsolete, you can find anything and everything related to accessible software applications, operating systems, and overall development for most disability communities at this site.

WebABLE! provides a similar resource. On the WebABLE! homepage, there is a direct link to Windows and Operating Systems (http://www.webable.com/op_systems).

Microsoft Windows 95/98/2000

The Microsoft Windows family of operating systems has achieved a significant level of accessibility for a commercial series of operating systems. There is still a long way to go, but Microsoft is committed to ensuring access to their products at every level. Microsoft's organization devoted exclusively to accessibility, the Accessibility and Disabilities group, maintains an up-to-date Web site keeping individuals informed of product accessibility.

Additionally, and again to their credit, Microsoft recently organized an advisory council (Microsoft Accessibility Advisory Council or MAAC) consisting of professionals and experts in the field of access technology to provide Microsoft with consultation and direction involving people with disabilities and software development. I have the privilege of serving on this council. To find out more about this effort, see http://www.microsoft.com/enable/news/aac0399.htm.

If you use Windows 95, 98 or 2000 (formerly Windows NT), I recommend that you visit Microsoft's Accessibility Web site (http://www.microsoft.com/enable/) to keep abreast of operating system developments.

III **10**

Developers and engineers should surf to http://www.microsoft.com/ enable/dev/default.htm for additional information regarding programming accessibility using the following Microsoft protocols:

Active Accessibility http://www.microsoft.com/enable/msaa/ default.htm

Screen Access Model http://www.microsoft.com/enable/msam/ default.htm

Java http://www.microsoft.com/enable/products/java.htm

Windows 95/98/2000 Accessibility Properties

Windows 95/98 and 2000 contain several accessibility features that you can easily access through the Windows Control Panel. To do this, complete the following steps:

1. From the Start menu, select Start>Settings>Control Panel.
2. Select the Accessibility Options icon. The Accessibility Properties window is displayed as shown in Figure 10.1.

Figure 10.1   Windows 98 Accessibility Properties window

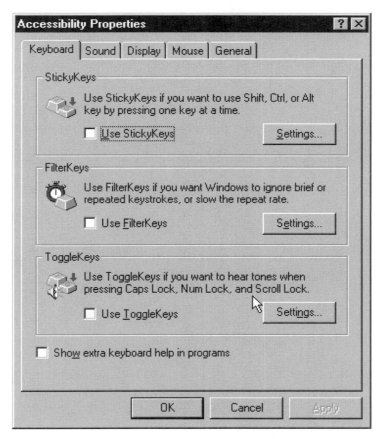

The Accessibility Properties window (also called the Accessibility Properties Sheet) includes a series of five tabs labeled Keyboard, Sound, Display, Mouse, and General. From this window, you can set most of the accessibility features available to the operating system.

Following is a list of the resident accessibility features currently available in Windows 95/98 and Windows NT Version 4.0 through the Accessibility Properties window. You can find a complete description of these features and other accessibility functions at http://www.microsoft.com/enable/products/winover.htm.

III   10

Keyboard

- StickyKeys enables you to set Shift, Ctrl, and Alt key combinations by pressing one key at a time.
- FilterKeys sets Windows to ignore repeated keystrokes or slows the repeat rate.
- ToggleKeys sets tones for the Caps Lock, Num Lock, and Scroll Lock keys.

Sound

- SoundSentry displays visual cues in place of (or in addition to) system sounds.
- ShowSounds enables an application that supports closed captioning to display captions in place of sounds.

Display

- High Contrast is an option you can set to specify colors and fonts that are designed for easier viewing.

Mouse

- MouseKeys enables you to control your mouse pointer using the keyboard's numeric keypad.

General

- SerialKey devices enable you to use an alternative device to specify keyboard and mouse commands.
- The General tab also enables you to set notification status and idle times for accessibility features.

Additional Features and Tools

In addition to the Accessibility Properties, the following features and tools are also available on all three platforms:

- Accessibility status bar indicators
- Adjustable cursor blink rate
- Hotkey combinations to set accessibility features or turn them on or off (see Table 10.1 for hotkey combinations)
- Support for DVORAK keyboard layout, which simplifies use for one-hand keystrokes

- Support for enlarged, solid, or inverted mouse pointers (inverted mouse pointers change color contrasts as they pass over your windows desktop and applications for easier viewing)

Table 10.1   Hotkey Combinations

Function	Hotkey
StickyKeys	Press the Shift key five times
FilterKeys	Press and hold the right Shift key for eight seconds
ToggleKeys	Press and hold the Num Lock key for five seconds
High Contrast	Press simultaneously left Alt key+left Shift key+Print Scrn
MouseKeys	Press simultaneously left Alt key+left Shift key+Num Lock

Windows 98 Enhancements

Windows 98 includes two new major features that enhance accessibility:

- The Accessibility Wizard, which makes it easier to customize your computer by grouping accessibility features
- Microsoft Magnifier (shown in Figure 10.2), a utility that enlarges portions of your screen in a separate window and enables you to adjust color schemes and contrasts

Figure 10.2   The Microsoft Magnifier

III   10

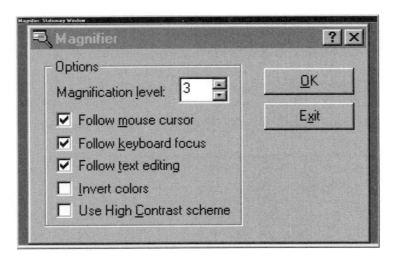

Windows 2000

Windows 2000 (formerly, Windows NT) contains several new accessibility features to assist users with disabilities. These features include the following:

On-Screen Keyboard, that displays a virtual keyboard on your computer screen making it easier for people with mobility disabilities to enter keyboard and keypad functions using joysticks or pointing devices.

Narrator, a text-to-speech utility that functions like a screen reader for blind and low vision users. Narrator works with Notepad, WordPad, Control Panel programs and Internet Explorer.

Utility Manager, which brings Magnifier, Narrator, and On-Screen Keyboard into a common area where you can easily launch any or all three of the applications.

Windows 2000 installs Narrator, On-Screen Keyboard, and Magnifier by default.

Windows 3.x/DOS

As you might expect, when an operating environment becomes nearly obsolete, support tends to lag. Windows 3.*x* and DOS are in that category. However, there are scores of freeware and shareware accessibility programs for both. In fact, I dare say there are more programs for both of these products than currently exist for the newer Microsoft Windows operating systems.

There are scores of Web sites (including Microsoft's) that provide downloadable shareware and freeware accessibility utilities. The most useful sites are as follows:

- Free and Cheap Windows Software for People with Disabilities
  `http://www.at-center.com/windows.html`
- Eye Tech Shareware Collection
  `http://www.deicke.org/etecshar.htm`
- Trace Research Microsoft Windows Toolkit
  `http://www.trace.wisc.edu/world/computer_access/win/winshare.html`

AccessDOS and Access Pack

Two products, same offering. AccessDOS was developed by Trace Research with support from IBM and the National Institute on Disability and Rehabilitation Research (NIDRR). Some of the functionality differs slightly from what is available in Windows 95/98, but in general, both AccessDOS and

Access Pack contain utilities that provide the same level of accessibility found in the Windows 95/98 Control Panel accessibility settings.

The AccessDOS and Access Pack features have become staples for most operating systems today. Their usefulness is immeasurable. Table 10.2 shows descriptions of these features.

Table 10.2   AccessDOS and Access Pack Features

Feature	Description
BounceKeys	Applications ignore unintentional keystrokes
MouseKeys	Mouse functions performed using keyboard
RepeatKeys	Adjust repeat rate of key when held down
SerialKeys	Use alternative input device in place of keyboard or mouse
SlowKeys	Applications ignore keystrokes that are held down beyond a defined period of time
SoundSentry	Set visual cues in place of system sounds
StickyKeys	Press one key to perform multikey functions
TimeOut	Stop AccessDOS after a defined period of time
ToggleKeys	Provide audio cues for Num Lock, Caps Lock, and Scroll Lock

UNIX/X Window and Linux

Few are aware that many accessibility tools exist for the UNIX platform. This tool development can be attributed, in part, to the pioneering work of the Disabilities Action Committee for X (DACX), whose goal is to develop accessibility features for workstations running X Window.

This volunteer group also created AccessX, a mirror package of the AccessPak utilities for UNIX-based windows systems. AccessX includes the following:

- StickyKeys
- MouseKeys
- RepeatKeys
- BounceKeys
- ToggleKeys
- SlowKeys

III   10

These utilities function as they do on other operating systems. AccessPak now ships with Sun Microsystem's Solaris and Digital Equipment Corporation's (now part of Compaq Computer) Digital UNIX operating systems.

Two additional shareware applications developed for the X Window platform are Puff and UnWindows. Puff is a screen magnification application that enables users to magnify both the text and graphic images on their computer screen. UnWindows also provides selective screen magnification and mouse pointer (cursor) tracking using visual and aural feedback mechanisms.

Recently several advancements in accessibility for Linux have been made, primarily focusing on access for the blind. A fantastic Web site monitoring Linux accessibility development is Hans Zoebelein's BLINUX Documentation and Development Project (http://leb.net/blinux/).

Please also refer to the section below discussing Emacspeak, an audio desktop for UNIX platforms.

OS/2 WARP

The IBM Special Needs Systems (SNS) group located in Austin, Texas has long been involved in the development of access systems for people with disabilities. They have not left their users without assistive aids and technology. While OS/2 WARP may not be as pervasive in the marketplace as Microsoft Windows, IBM still maintains a solid user base. OS/2 Warp includes special needs keyboard settings, including the following:

- StickyKeys
- Acceptance Delay
- Delay before Repeat
- Repeat Rate

You can refer to the SNS home page (http://www.austin.ibm.com/sns/index.html) for additional information about OS/2 accessibility and other accessibility technologies IBM provides.

Mac OS

At one time, Apple and IBM ruled the world of accessibility for people with disabilities. At the very least, both corporations deserve recognition as pioneers in the advancement of accessibility. Recently, both corporations have renewed their commitment to accessibility issues, and Apple continues to provide excellent customer service and outstanding support to children and students with disabilities.

Every Macintosh computer includes the following set of tools:

- Easy Access utilities (StickyKeys, SlowKeys, and MouseKeys)
- CloseView screen magnification software
- Key-repeat disable
- PlainTalk (text-to-speech synthesis and voice recognition)
- StickyMouse
- Visual alert cues for the deaf and hard of hearing

Beginning with Mac OS 8 operating system, all of these tools have been packaged as part of the Universal Access tools and can be installed from the Mac OS 8 installation software (select Item 10 on the installation CD-ROM). For Mac OS 8.5, double-click the CD Extras folder, and then double-click the Universal Access Folder icon. Drag the CloseView and/or Easy Access icons on top of your system folder to install.

Brief descriptions of the major utilities follow.

Keyboard Options

StickyKeys enables you to press one key at a time for any command that requires simultaneously pressing two or more keys. Beginning with Mac OS 8, the new version of Easy Access enables Sticky Keys to remain active after waking up if the feature was active when your PowerBook or Portable shut down.

Some individuals with motor or mobility disabilities have difficulty accurately typing keys on a keyboard. Slow Keys enables you to delay the keyboard's response to a typed key should you accidentally type the wrong key.

Keyboard control panel sets keyboard repeat speed (and delay until repeat) as well as choice of 20 language layouts.

Sound

Talking Alerts enables text-to-speech software to read the text of onscreen alerts after a user-defined period of time. However, Talking Alerts only functions on modal alert messages. You activate Talking Alerts through the Speech Control Panel.

III 10

Display

CloseView is a screen magnification tool for low vision users. CloseView enables you to enlarge (magnify) screen contents up to 16 times.

Mouse

MouseKeys enables you to control your mouse pointer using the keyboard's numeric keypad. Because the PowerBook does not include a numeric keypad, Apple includes this special utility, which emulates the numeric keyboard and then functions as the standard MouseKeys utility.

Mouse Control Panels enables you to set the speed of mouse tracking and double-click speed.

Easy Access/Universal Access now remembers whether StickyKeys, MouseKeys, and SlowKeys were on or off between restarts.

For additional information about accessibility applications and internal tools available to the Mac system, surf to `http://www.apple.com/education/k12/disability/`. For additional shareware — from changing the size of the caret and arrow to onscreen keyboards — visit Trace's Web site at `http://www.trace.wisc.edu/world/computer_access/mac/mac-share.html`.

Screen Readers

One of the most significant advances in the field of information technology for the blind community involves screen reader technology. Screen readers are software applications combined with a synthetic voice that reads computer data back to the blind user. This includes all screen objects (for example, windows and icons). Screen reader technology makes it possible for people with visual disabilities to use operating systems built upon the graphical user interface (GUI).

The objective of the screen reader today is to efficiently render graphical entities displayed by an operating system. This is much harder than it may sound. Ultimately, the technology must track every system and application event in addition to the standard user action. In order to do this, most screen readers employ an *off-screen model* technology — more or less a shadowing protocol that tracks, intercepts, converts, and then renders the event or user action using synthetic speech output. The description alone should be an indication of how many additional computing cycles are required to maintain this user environment. You can now envision how hard it is for a person using a screen reader to follow (and comprehend) simultaneous programming events, such as Web frame technology.

Recently, however, several advancements in screen reader technology enable users to interact more easily with Web browsers, thus making it possible to render Web content accessibly. Most screen readers can now access and efficiently render Web pages, even those containing complicated structures including forms, frames, and tables.

In particular, great strides have been made to deal with Web pages designed with tables. As mentioned in previous chapters, a major issue involving difficult access to Web pages was the inability for a blind person using a screen reader to easily navigate a Web page constructed with tables for presentation purposes.

Traditionally, a screen reader reads data in a left-to-right manner. When it reaches the end of a perceived line, it goes on to the next line. However, when text is columnar in nature, the screen reader renders data across the same line, regardless of the column break.

Because a Web page often consists of combinations of text, objects, multimedia, and images, navigating a Web page can be a nightmare. Screen readers now take various approaches to rendering a Web page properly, including reformatting the page into a single column that can then be easily read by the application using a voice synthesizer.

As a result, there may be less of a need to develop specialized Web browsers (as discussed in Chapter 7). Most of the popular screen readers work with traditional Web browsers, including Internet Explorer, Navigator, and Lynx. However, much like operating systems and word processors, preferences abound. I recently conducted an informal survey to determine which screen readers were preferred when used to access the Web. Not one stood out head and shoulders above the others. It was clear that each user had his or her own personal preference.

Systems administrators and Web site developers should consider testing their Web sites using screen reader technology. At the very least, use one of the self-voicing browsers to get some idea of how a Web page sounds to a visually disabled user. This might seem tedious, but keep in mind that voice technology is slowly making its way into mainstream technology. Traditional graphical browsers available on computers will soon give way to voice-based user agents. Getting a jump on the technology today will go a long way toward development tomorrow.

Most screen reader vendors offer free trial versions of their software through Web sites. The next few sections provide brief descriptions of the more commonly used screen reader applications, the client and operating

III 10

system platforms they support, and associated URLs. All of these screen readers can be used by a blind or a person with a visual disability to access the Web.

Screen Reader/2

Screen Reader/2 is an IBM product that works on both OS/2 WARP and Windows platforms. Screen Reader/2 has recently been updated to support Netscape Navigator. However, Screen Reader/2 is particularly well suited for IBM's Web browser, IBM Web Explorer. IBM provides complete support, including a newsletter for Screen Reader/2, at `http://www.aus-tin.ibm.com/sns/snssrd.html`.

Slimware Windows Bridge

Produced by Syntha-Voice Computers of Canada, Slimware Windows Bridge is a screen reader that provides speech and Braille access to Windows. Bridge includes support for Microsoft Internet Explorer 3.0 and 4.0 as well as Netscape Navigator 4.0.

Slimware Windows Bridge is available for Windows 98/95, 3.1, and MS DOS. Windows Bridge reads Web pages in a natural order and does not rearrange or reformat a Web page (as some screen readers do). You can download a free demo of Windows Bridge at `http://www.syn-thavoice.on.ca/`.

Note

To run the Windows Bridge demo, you must have a sound card installed.

Window Eyes

Window Eyes, a product developed by GW Micro, is a screen reader known to work well with Netscape Navigator. However, Window Eyes is also compatible with Internet Explorer and includes support for Microsoft Active Accessibility (MSAA). This is a key attribute because Microsoft uses MSAA to expose more of the interface and internal information to screen readers (as well as other assistive technologies).

Window Eyes is supported on Windows 3.11/95/98. GW Micro includes DOS support using their Vocal Eyes screen reader.

You can download the Window Eyes trial demo by following the instructions at `http://www.gwmicro.com/gwtext/download_text.htm`.

JAWS

Job Access for Windows Software Version 3.2 (JAWS), another popular windows screen reader application, includes support for Windows 95/98 and NT. JAWS is a product of Henter-Joyce, Inc.

JAWS is the kind of screen reader that reformats a Web page. Many users find this attribute appealing. For example, JAWS reformats list links alphabetically and posts the results in a list box. Subsequently, the user can then easily scroll through each link. JAWS also reformats tables using a simple function key (F5) operation. The table is converted into a single column of text including the appropriate column headers.

JAWS is compatible with both Internet Explorer and Netscape Navigator. You can download a 40-minute demonstration version of JAWS at `http://www.hj.com/JFW/JFW.html`.

WinVision 97

Artic Technologies' screen reader product, WinVision 97, is an MSAA-compatible application that works optimally with Internet Explorer 3.2. However, WinVision also works with current versions of Internet Explorer and Netscape Navigator.

WinVision is also Microsoft Office 97–compatible, a highly valued characteristic for many users. WinVision 97 is available for Windows 3.1 and 95.

You can download a 30-day trial version of WinVision 97 at `http://www.artictech.com/whywv97.htm#top`.

III 10

outSPOKEN Solo

outSPOKEN Solo 2.0, a product of the Holland-based corporation ALVA, is a screen-reading program for Windows 98 and Windows 95. Originally a product designed and developed by Berkeley Access of Berkeley Systems (well known for their screen saver products), outSPOKEN was the first publicly-released windows screen reader available for the Apple Macintosh personal computer.

outSPOKEN Solo 2.0 is compatible with Microsoft Internet Explorer 3.0 and 4.01 as well as Netscape Navigator 3.0 and 4.0.

outSPOKEN 1.21 is available for Microsoft Windows 3.1, and outSPO-KEN for Macintosh supports Mac OS 8.

You can download all of the outSPOKEN demos from `http://www.aagi.com/`.

HAL

Dolphin Computer Access Limited, located in the United Kingdom, is well known for their screen reader, HAL. HAL can be installed on both Windows 95 and 98 as well as Windows NT Workstation. You can use HAL with Netscape Navigator and Internet Explorer.

HAL (similar to other screen readers) supports several languages including British and American English, Spanish, Greek, and Swedish.

You can download HAL at `http://www.dolphinuk.co.uk/demos/download.htm`.

UltraSonix

UltraSonix is the name of the screen reader that evolved out of the Mercator Project at Georgia Tech. The goal of this project is to make UltraSonix available to visually impaired users running X Window system applications on personal computers using Linux.

For more information about how to download UltraSonix for Linux, go to `http://www.trace.wisc.edu/world/computer_access/pusl/`.

ASAW

MicroTalk's Automatic Screen Access for Windows (ASAW) is a Microsoft Windows–compatible screen reader. ASAW supports Windows 3.1, Windows for Workgroups, and Windows 95. Windows 98 and Windows NT versions are in development.

You can find more information about ASAW at `http://www.microtalk.com/asawinfo.html`.

Not a Screen Reader, But . . .

SurfTalk for the Macintosh — used in conjunction with Netscape Navigator or Internet Explorer — enables the user to speak any visible hyperlink as well as hyperlinks saved as bookmarks. SurfTalk enables the user to speak common navigational commands such as "go back," "reload," and "add bookmark." Demo available. Visit `http://www.surftalk.com/surftalk/index.html` for more information.

Multimedia Tools

My greatest fear about the accessibility of the World Wide Web today surprisingly does not involve access for the blind and people with visual disabilities — although they are easily the disability community most affected by Web accessibility barriers today.

Rather, my greatest fear involves access for the deaf and people with hearing disabilities. The person who is not able to speak or the person who has difficulty speaking are similarly affected. Emerging Web and computing applications are focusing on interfaces that require more of our hearing and speaking senses — the two senses these communities are without.

The good news is that several companies and nonprofits, including the W3C and the National Center for Accessible Media (NCAM), have spearheaded projects and enhanced products to make multimedia more accessible, particularly for the deaf. This includes captioning and descriptions for streaming audio and video.

Web access systems for the deaf and people who have hearing disabilities usually involve a captioning system or application combined with a "player" or plug-in that can render the captions. All three of the popular streaming media players support captioning for the deaf. They include:

- Microsoft's Windows Media Player
- RealNetworks RealPlayer
- Apple's QuickTime Player

Displaying captioning in each of the three players is a fairly simple task. Following are the instructions for each player.

Windows Media Player

To display captions using the Windows Media Player, do the following:

- Click on the View menu
- Select the Caption menu item

See Figure 10.3 to view an example of the Windows Media Player displaying captions.

RealNetworks RealPlayer

Note that there are several RealNetworks players available on the market today. However, the process for displaying captions is consistent within each player. To display captions, do the following:

- Click on the View menu
- Select the Preferences... menu item. This displays the Preferences property sheet.
- Select the Content tab
- Click on the Settings... button in the Accessibility property area. This displays the Accessibility Settings dialog box.
- Click on the Show Captions radio button.

Alternatively, you can click on the Use Accessibility features when available checkbox. This displays captioning and/or descriptive audio tracks when the media includes it.

QuickTime Player

The QuickTime Player is now available on both the Mac and Microsoft Windows platforms (for that matter, RealPlayer and Windows Media Player are also available for Mac OS). To display captioning on the QuickTime Player, do the following:

- Click on the Edit menu.
- Select the Enable Tracks... menu item
- Click on the Text Track entry to turn on text description

Naturally, the key to playing accessible content, is to ensure that the media itself contains accessible coding. They most popular of these media types are:

- W3C's Synchronized Multimedia Integration Language (SMIL)
- Microsoft's Synchronized Accessible Media Interchange (SAMI)
- Apple's QuickTime

SMIL

Early in 1998, the W3C released the Synchronized Multimedia Integration Language (SMIL 1.0, pronounced "smile") specification. It enables Web developers to create multimedia presentations made up of pictures, sound, and text contained as separate elements and then synchronized when the player accesses the appropriate hyperlink. The same document can also include streaming audio or video. All of it can be coded using a standard text editor. If you can code an HTML page, you can code SMIL.

The complete SMIL specification as well as examples of SMIL code can be found on the W3C's web site at http://www.w3.org/TR/REC-smil/. The W3C also has a web page devoted to the accessibility features of SMIL at http://www.w3.org/TR/SMIL-access. Lastly, I recommend that you review Larry Bouthiller's SMIL tutorial that appears on the Web Techniques E-Zine site at http://www.webtechniques.com/archives/1998/09/bouthiller/.

SAMI

Microsoft's Synchronized Accessible Media Interchange (SAMI) enables you to create Web pages and multimedia that include closed captioning for users who are deaf or having hearing disabilities. SAMI has its own file format that Microsoft offers as an open standard (that is, you don't pay licensing fees). You can use Microsoft's Windows Media Player to view the source file. You can install the Windows Media Player on Windows 95/98 or 2000 systems.

The beauty of creating a SAMI document is that it is constructed like an HTML document. For example, note the sample code that Microsoft provides with the SAMI (.asf file extension) download:

```
<SAMI>
<Head>
<Title>President John F. Kennedy Speech</Title>

<STYLE TYPE="text/css">
<!--
P {margin-left: 4pt; margin-right: 4pt;
font-family: sans-serif; font-size: 14pt;
text-align: left; font-weight: normal;
color: white; background-color: black;}

.ENUSCC {Name: English; lang: en-US;}

#Source {margin-bottom: -12pt; margin-left: 0pt;
margin-right: 0pt; padding: 2pt;
background-color: silver; color:
```

III 10

```
black; font-size: 10pt; font-weight: bold;
font-family: sans-serif;}
-->
</Style>

</Head>

<Body>

<SYNC Start=250>
<P Class=ENUSCC ID=Source>Pres. John F. Kennedy
<P Class=ENUSCC>Let the word go forth, from this
time and place to friend and foe alike.

<SYNC Start=7200>
<P Class=ENUSCC>That the torch has been passed,
to a new generation of Americans,

<SYNC Start=13500>
<P Class=ENUSCC>born in this century, tempered by war,
disciplined by a hard and bitter peace,

<SYNC Start=23000>
<P Class=ENUSCC>proud of our ancient heritage, and
unwilling to witness or permit

<SYNC Start=30000>
<P Class=ENUSCC>the slow undoing of those human rights

<SYNC Start=36000>
<P Class=ENUSCC>to which this nation has always been
committed

<SYNC Start=40000>
<P Class=ENUSCC>and to which we are committed today at
home and around the world.
```

```
<SYNC Start=46000>
<P Class=ENUSCC ID=Source> 
<P Class=ENUSCC> 

</Body>

</SAMI>
```

Figure 10.3   Windows Media Player rendering a SAMI file with closed captioning embedded in an HTML Web page

Microsoft's Accessibility group provides an informative Web site about SAMI at `http://www.microsoft.com/enable/sami/default.htm`.

Other Players

Players often function both as standalone multimedia browsers or plug-ins to Web browsers, including Netscape Navigator and Internet Explorer. The following players all support captioning for the deaf and are SMIL-compatible.

- NIST's S2M2 SMIL player
  `http://mpeg4.nist.gov/player`
- The Productivity Works LpPlayer
  `http://www.prodworks.com/productsindex.htm#lpplayer`
- SIGTUNA Player
  `http://www.jsrd.or.jp/dinf_us/software/software.htm`
- Apple's QuickTime player
  `http://www.apple.com/quicktime/`
- Compaq's Hypermedia Presentation and Authoring System (HPAS)
  `http://www.research.digital.com/SRC/HPAS`

ShockTalk (`http://www.surftalk.com/shocktalk/index.html`) is a Macromedia Xtra (plug-in) that enables Director developers to create products or rapid prototypes with a spoken interface. This is a cross-platform interface for speech recognition and speech synthesis technologies. It supports Director applications for the desktop and Shockwave movies for the Net. ShockTalk is being developed by Digital Dreams to facilitate the design and application of speech-centered user interfaces.

Tools for Creating Accessible Web Media

Several applications and publishing suites exist to help you create accessible multimedia, streaming audio and video. Following is brief list of those currently available.

MAGpie

WGBH's National Center for Accessible Media (NCAM), the recognized experts in the area of accessible media recently released the first version of their Media Access Generator or MAGpie (`http://www.wgbh.org/pages/ncam/webaccess/magindex.html`). MAGpie allows you to add captions to all three multimedia formats including QuickTime, SMIL, and SAMI.

MAGpie also features the ability to integrate audio descriptions into SMIL presentations. MAGpie is a free download.

QuickTime Pro

You can create SMIL-compatible, multimedia presentations for the Web that support text captioning on the Web with Apple QuickTime 3.0 or later. (http://www.apple.com/quicktime/). QuickTime supports digital video formats like AVI, AVR, DV, and OpenDML.

LpStudio Pro

The Productivity Works' LpStudio Pro (http://www.prodworks.com/productsindex.htm#lpstudiopro) is a software system that provides professional capabilities to create, edit, and produce multimedia materials including audio, text, and images. The system is specifically designed for spoken audio as a key component of the resulting materials, and it embodies technology that enables editing by both spoken phrase and digital waveform. LpStudio Pro supports production of HTML, XML, and SMIL standards from W3C.

Additional SMIL supported authoring tools include:

- Allaire's HomeSite
  http://www.allaire.com/Products/HomeSite/
- Sausage Software's SMIL Composer SuperTool
  http://www.sausage.com/supertoolz/toolz/stsmil.html
- ConFluent Technologie's Fluition (Macs only)
  http://www.electricleisureland.com/cgi-bin/cti.cgi

If you are interested in validating your SMIL code, a web validation service can be found at http://dejavu.cs.vu.nl/~symm/validator/).

Closing Thoughts

Accessible multimedia doesn't create itself. Don't let anyone fool you – it is time consuming and meticulous work, particularly if the captioning or descriptive text is not included as part of the original production. Consequently, very few Web media networks produce accessible multimedia; fewer still believe it's an important part of their business.

While this is certainly discouraging, an opportunity for innovation truly exists. Prior to publishing this book, I came in contact with a company who has developed a means for generating on-the-fly captioned streaming media using voice-recognition technology. As an application service provider

III **10**

(ASP), the DashCenter (http://www.dashcenter.com/) uses the Virage (http://www.virage.com/) video application server software to produce accessible news and entertainment productions of streaming audio and video. This is the wave of the future that I look forward to.

Other Useful Web Accessibility Services

You knew it wouldn't be long before commerce had its way with the Web. Today's e-commerce uses software applications that emulate traditional customer services and product sales techniques. Banking, shopping, stock trading, and software updates are just a few of the services you can now access from the convenience of your Web browser.

As part of this growth spurt, several Web sites are now offering assistive services to people with disabilities. The following Web services are freely available to all users.

Access Adobe

Access Adobe (http://access.adobe.com/) was the first publicly available Web accessibility service (other than the Bobby validation service) created specifically to enhance the Web experience for users with visual disabilities.

Access Adobe enables visually disabled Web surfers to read Adobe Page Description Format (PDF) files. PDF files in their native state are read using the Adobe Acrobat Reader and are inaccessible to the people with visual disabilities because PDF is a graphical output file type that cannot be read by a screen reader. It is a popular file output type, particularly for those involved in desktop publishing and publishers of large electronic documents. The United States Internal Revenue Service (IRS) is a huge publisher of PDF documents.

The Access Adobe service converts PDF documents into HTML or ASCII text that can then be read screen readers using speech synthesizers to output speech. Access Adobe provides three ways to convert a PDF document:

- Form submission
  http://access.adobe.com/access_form.html
- E-mail
  http://access.adobe.com/access_email.html
- Acrobat plug-in
  http://access.adobe.com/access_plugin.html

To use the form submission service, simply enter the same URL of the source PDF document. The resulting document is converted by the Access Adobe service.

Access Adobe is a useful tool, particularly for smaller documents. Some members of the blind community are bothered by inconsistencies in the transformation process, particularly for larger document conversions. The service also has some difficulty in properly rendering columnar text. However, Adobe is committed to accessibility for people with disabilities in all their products, including Acrobat, PDF, PageMaker, GoLive, and their other mainstream publishing tools. Adobe and Microsoft are working together to support the Microsoft® Active Accessibility (MSAA) Application Programming Interface (API) For additional information, refer to: `http://www.adobe.com/aboutadobe/pressroom/pressreleases/200004/20000418acr.html`.

The Web Access Gateway

The Web Access Gateway intercepts your browser's request to access a Web site and then converts that Web site's page to a more easily readable rendition for screen readers and text browsers.

Using a selection process, the Web Access Gateway enables you to choose from a variety of settings that enhance the accessibility of a Web site. Once you have made your selections, you enter the URL of the Web site you want to visit. The Gateway then retrieves the Web page and renders it based on the settings you specified. Using the Gateway, you can specify changes including color, font size, frame and table renderings, and wrap text links.

The Web Access Gateway's creator, Silas Brown, encourages others to download the source code. You can find more information about The Web Access Gateway at `http://members.bigfoot.com/~silasbrown/access.html`.

III **10**

Betsie

The BBC Education Text to Speech Internet Enhancer (Betsie), like the Web Access Gateway, is one of the most innovative Web-based accessibility services available today. Offered by the British Broadcasting Corporation (BBC), Betsie takes a Web page, strips out complex HTML elements, and converts the page into an easy-to-read text version of the page.

At first glance, it appears that Betsie is just a text version of the complete BBC Web site. In fact, once you access the Betsie home page, every BBC link followed leads to a page that is converted on the fly. For example, a graphic version of the BBC Web page appears in Figure 10.4.

Figure 10.4 Graphic version of the BBC home page

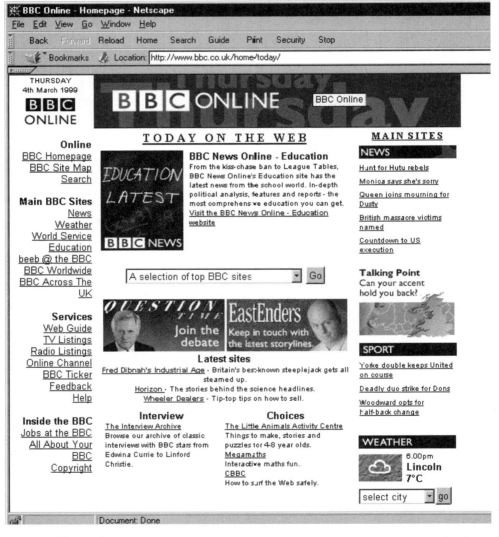

Using Betsie, the page is converted to a text version as displayed in Figure 10.5.

Figure 10.5   Betsie version of the BBC home page

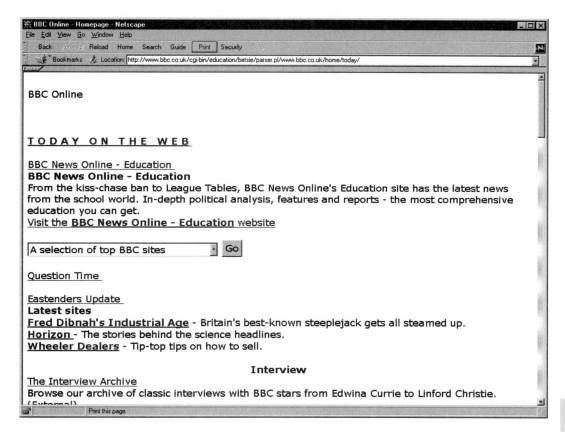

The BBC makes the Betsie source code available to anyone interested in offering the same service from his or her Web site. You can download the source code from `http://www.bbc.co.uk/education/betsie/tech.html`.

To learn more about how Betsie works, refer to the complete technical description at `http://www.bbc.co.uk/education/betsie`.

Emacspeak

Emacspeak is not a Web accessibility service; it is a complete speech output system or audio desktop. Without a doubt, it is one of the greatest applications available to blind and users with visual disabilities everywhere. Emacspeak was created by world-renowned research scientist and my very good friend, Dr. T.V. Raman. Raman provided the following description of the product:

> Emacspeak uses InterActive Accessibility technology (IAA) to provide a powerful, Internet-ready audio desktop, that completely integrates Internet technologies, including Web surfing and messaging, into all aspects of the electronic desktop. The result is a fully functional audio desktop that provides complete eyes-free access to all major open 32 and 64 bit operating systems. By seamlessly blending all aspects of the Internet such as Web-surfing and electronic messaging into the audio desktop, Emacspeak enables speech access to local and remote information with a consistent and well-integrated user interface.

Unlike screen readers, which rely on offscreen models and a significant amount of programming overhead (read: memory hog), Emacspeak provides spoken feedback for everything you do once the Emacs editor is loaded. Because Emacs is incredibly extensible, there isn't a task that you cannot perform within the editing interface, including Web browsing. Using Bill Perry's powerful and fully HTML 4–compliant W3 browser (discussed in Chapter 7), the Web becomes completely accessible to the blind user. Emacspeak-98 contains additional support for Aural Cascading Style Sheets (ACSS), a part of the W3C's CSS2 specification.

Emacspeak is available at `http://www.cs.cornell.edu/home/raman/emacspeak/emacspeak.html`.

Summary

In this chapter, you learned about specialized access systems used by people with disabilities. These access systems enhance the Web experience in the following ways:

- Provide integrated operating systems utilities to facilitate easier use of computer and system software.

- Blind and users with visual disabilities can access Web pages through voice and Braille output with screen reading software. Most screen readers can read render Web pages in conjunction with standard Web browsers, including Internet Explorer and Netscape Navigator.

- Include captioning and text descriptions of multimedia content. Using the SMIL standard or SAMI file output, authoring tools and players can easily render captioning for the deaf.

- Take advantage of several free Web-based services that perform valuable functions, such as converting PDF to HTML, and automatically converting complex Web pages into easy-to-navigate Web sites and complete audio desktops.

References

Trace Research. *Designing a More Usable World For All — Computers and Software.* Available from `http://www.trace.wisc.edu/world/ computer_access/`.

EmacSpeak-98. (Created by Dr. T.V. Raman, 1998.) Available from `http:// cs.cornell.edu/home/raman/emacspeak/emacspeak.html`.

Microsoft Corporation. Synchronized Accessible Media Interchange (SAMI) Web page. Available from `http://www.microsoft.com/enable/sami/ details.htm`.

III 10

Chapter 11

Looking Ahead Towards Emerging Technology

In This Chapter

- Enabling technology: Is there a future?
- The National Technology Grid
- Personal computing
- The Web and TV
- Software architectures
- Achieving pervasive accessibility

The issues of next-generation interfaces, globalization, personalization, information dissemination, and converging emergent technologies have created an incredible dichotomy for people with disabilities: the widespread proliferation of multimodal information through advanced network systems, disseminated anywhere, anytime — yet most designed and developed with little thought regarding accessibility to people with disabilities.

This is a vicious cycle that people with disabilities know all too well. With each new technology, the disabilities community often is offered nothing more than quick-fix solutions to accessibility problems. Subsequently, technology that increases access to people with disabilities is termed *assistive* and *adaptive* as opposed to being a technology that is inherently *accessible*.

The focus of this chapter is to examine a few emerging information technologies and how each implements measures to advance Internet and Web accessibility for people with disabilities.

Enabling Technology: Is There a Future?

In an age when technology buzzwords such as *immersive* and *Universal Design* imply access for all, you may wonder why emerging technology isn't designed accessibly from the start. Certainly, advances in human-centered design methodology and processes promote usability for every user, right? In theory yes, in practice no.

User-centered design processes and methods are sound. The problem lies in practice. The level of user participation involving people with disabilities is practically nonexistent. Consequently, next-generation interfaces are rarely prepared to deal with the inevitable — the moment the interface goes public, millions of people with disabilities are not able to use it.

Fortunately, not all is lost. While much of advanced technology is sorely lacking in accessibility (and usability, for that matter), recent advances promoting accessibility give rise to hope for the ultimate technological goal: pervasive accessibility. In other words, interfaces that adapt to the user, regardless of ability.

The National Technology Grid

The new millennium Web is headed away from its current paper and document paradigm to a programmable, software-engineered user interface. Information technology is moving toward an embodiment of multimedia: electronic libraries; digital audio books; streaming video; streaming audio; application-designed Web pages presented in virtual reality; teleimmersive, collaborative environments accessed using simple desktop systems; neural interfaces; intranets; extranets; wireless communications; and a plethora of other telecommunication entities.

The U.S.-based National Computational Science Alliance (NCSA) describes the new paradigm simply as the National Technology Grid. No doubt, this is the first step toward an international grid. It is the essence of globalization, the power driving both the National Information Infrastructure (NII) and Global Information Infrastructure (GII).

But Will It Be Accessible?

To the credit of NCSA (`http://ncsa.uiuc.edu/`), making the National Technology Grid accessible is a goal. NCSA Director Larry Smarr stated, "We will design the Alliance and the Grid in a manner that is accessible to all. Moreover, the Alliance is committed to increasing participation of groups (particularly women and minorities) and institutions historically underrepresented in advanced computation."

Education, Outreach, and Training

An integral part of the project involves the Education, Outreach, and Training (EOT) activity jointly supported by NCSA and National Science Foundation Partnership for Advanced Computational Infrastructure (NPACI). A key member of the EOT is the Trace Research and Development Center at the University of Wisconsin at Madison. As mentioned throughout this book, Trace is the leading U.S.-based resource for accessibility and information technology.

The thrust of the EOT is to build awareness about the resulting Alliance technologies, applications, and materials for the general user community, including children, women, people with disabilities, and other minorities.

The Universal Design/Disability Access Program

Trace subsequently created the Universal Design/Disability Access Program. This program is designed to ensure that next-generation computational science and advanced computing environments are usable and accessible to all people.

Through this project, five key activities have been initiated:

1. Build general awareness of universal design and accessibility for people with disabilities.
2. Provide specific materials and presentations to support NCSA and PACI.
3. Create an associated resource Web site.
4. Produce a Java developer set of guidelines and toolbook for the collaboration tools group.
5. Develop a Web accessibility toolkit.

III

11

Personal Computing

Personal computing has taken enormous leaps in the past five years. In that time, we've seen the average new desktop computer change from a 486-based, 66MHz, 8MB memory, 14-inch monitor, 9600bps modem configuration to a Pentium III-based, 500MHz, 128MB memory, 21-inch monitor, 56K modem configuration. In spite of these advances, the most intriguing addition to the PC has nothing to do with performance and everything to do with the user interface: voice recognition (also called spoken language systems) and speech output systems.

The irony of the voice recognition and speech output paradigm shift is that they are not new technologies. On the contrary, the blind and physically challenged communities have used voice recognition and speech output systems for many years. Other disability communities including the nonverbal, people with hearing disabilities and people with cognitive disabilities also use both technologies. Companies such as Dragon Systems (http://www.dragonsys.com/), Kurzweil (http://www.lhs.com/education/), and IBM (http://www.ibm.com/) were developing these interfaces as commercial products for the disabilities communities back in the mid-eighties.

Note

Prior to publication, Kurzweil Educational Systems and Dragon Systems were sold to Lernout and Hauspie Speech Products.

Support for Spoken Language Systems

To increase the development of this user interface, companies including Microsoft Corporation, IBM, and Sun Microsystems are making sure that operating systems and applications such as Windows and Java can handle the application requests. Recently, IBM released the first version of their Java-based, self-voicing kit (SVK), which enables automatic speaking of Java applications. Chip manufacturers such as Intel are creating 64-bit processing chips to handle the kind of computing cycles required to generate much more complex spoken language interfaces.

Advances similar to spoken language systems are terrific for many communities of people with disabilities. However, every now and then, advanced development of an accessible technology such as voice recognition acts as a double-edged sword — it cuts through the barriers presented to one type of disability and cuts out the accessibility for another.

Potential Conflicts

The excitement generated by voice recognition has PC manufacturers and software companies working furiously toward the goal of a total voice-based system. The same technology is being embedded into Web and Internet applications.

However, consider this: If a system's input and output technology is solely based on voice input/output (I/O), how will people who are deaf use the system? How will they hear the voice output? And what will they do if they are unable to speak? Even people who do not have a disability could be affected. For example, how does a voice-based interface distinguish and properly interpret commands in an office environment plagued by noise pollution?

Similarly, there are individuals who are nonverbal or have a disability that otherwise does not allow them to speak easily or clearly. This is often the case of a person with advanced multiple sclerosis, Parkinson's disease, or Lou Gehrig's disease (ALS). So, while advances in personal computing technology may appear to be achieving new levels of accessibility, as noted earlier, these interfaces must include participatory design by people with disabilities to ensure that they are truly user-centered. Additionally, while the "Star Trek" interface may appear appealing, it has to be acknowledged that an interface strictly based on voice is not a universal solution.

Additional PC Advances in Accessibility

With regard to future personal computing advances, you can look forward to the following:

- Universal Serial Bus (USB) support for all assistive and adaptive technology.
- Neural interfaces using brain and sensory technology that enable an individual to interact with a computer. A great example of this technology is the CyberLink headband. This headband enables people who have no use of their hands, or who are perhaps paralyzed, to interact with their computer by sensing head and brain signals and translating those to computer input actions. Eye-gaze and eye-tracking interfaces provide similar functionality.

III

11

- Advances in infrared technology enabling devices and information appliances to more easily facilitate advanced technology use by people with disabilities (for example, wireless communications, personal computing, navigation systems, and consumer electronics).

The Web and TV

Web commerce is moving ahead at breakneck speed. Almost with no regard for success or failure, Web-based products, services, and initiatives are appearing everywhere. Talk about pervasive — on an average day I must receive a dozen e-mail messages advertising a new Web entity. Imagine what venture capitalists see in a day!

In the effort to attract more users and consumers to the Web, there are two key delivery mediums: citizen access points and common user interfaces. Citizen access points include ATMs, public telephones, and kiosks. Common user interfaces include televisions, telephones, appliances, and (now) personal computers.

The convergence of common user interfaces along with their associated networks and communication protocols is fascinating, to say the least. It has promoted the move toward products that display (or otherwise render) multiple data types using multiple views by way of a single display device. One such technology that I believe will have certain success, particularly for people with disabilities, is the convergence of television and the Web.

The Television

The combination of television and the Web is priceless. Bundle them with cable or satellite network services and you've got gold. Many consumers find the personal computer difficult to use. Many others feel it's not worth the expense. But there are very few households that don't have at least one television. The television is a simplified network-enabled device that's easy to use and generally accessible. Those proclaiming the death of network computers (NCs) better think twice. It's been sitting in their living room all along. Better still, this device includes built-in captioning (mandated by U.S. law) and support for descriptive video services through the SAP channel. (Descriptive video is a specialized narrated video service for the blind.)

It's true that screen resolution is still poor and does not match that of most computer monitors, but advances in digital television and next generation display monitors will likely improve readability. Portability better suited through the cell phone and personal digital assistant (PDA), but I'll bet dollars of donuts most people will still do their web surfing using the "tube"!

The Combined Web and Television Interface

WebTV is a combination of a Web service and receiver that enables you to view television programming and Web pages, simultaneously if you want. Along with the service and receiver, you use the associated remote control or separately purchased keyboard to navigate. The power of the Web in a television is the ideal network computer and information appliance.

Note

It should be noted that WebTV is a brand name registered by WebTV Networks, now a subsidiary of Microsoft Corporation. It is not the only product developed that combines the Web and television interfaces. For example, TV onthe Web is another available service (http://www.tvontheweb.com/).

Consider the possibilities of a combined Web and television interface within the educational field alone. Often times, due to circumstances beyond their control, children with disabilities are unable to maintain a standard school schedule. This could be due to hospitalization, long and intense rehabilitation, lack of accommodation, or other limitations. The point is, the typical educational process is delayed, truncated or otherwise aborted.

Now consider the potential solution. Hospitals, rehabilitation facilities, and homes generally contain or have access to a television. Televisions linked to the Web could easily be configured to accommodate students, helping them continue their education using Webcasts of classes, access to a plethora of research material, and Web-based tutorials and educational systems. Combined with collaboration tools, there is virtually no limit to one's ability to continue the educational process with almost no interruption and at an affordable cost.

WebTV is still a relatively untried technology where people with disabilities are concerned. Accessibility issues involving the keyboard and the ability of the service to deliver accessible content present interesting challenges for the future. Microsoft, the parent company of WebTV Networks, has already assigned a member of their accessibility development group to work

III

11

on improving the technology's accessibility. The CPB/WGBH National Center for Accessibility (NCAM) is also working on captioning and other accessible improvements for WebTV.

Software Architectures

The technology wave will go as software goes. The accessibility of the Web, an interface driven by software development, is no different. Recent advancements made at Sun and Microsoft involving two key software components, the Java Accessibility and the Microsoft Active Accessibility (MSAA) architectures and interface standards, ensure greater levels of accessibility for people with disabilities in the coming years.

Both architectures not only increase the accessibility of the operating environments distributed by Sun and Microsoft, but they encourage the development of accessibility at the application and operating system levels for all software development organizations. All of the entities, tools, hooks, utilities, and enabling coding constructs are available on a system-wide basis. This is not a fluke, but rather a key strategy implemented and promoted by both corporations to ensure that application developers can develop accessible next-generation interfaces through the next millennium. In other words, the infrastructure is there; you just need to take advantage of it.

Cross-Reference

Please refer to Chapter 9 for a detailed discussion of Java Accessibility programming practices.

Brief descriptions of both the Microsoft and Sun accessibility platforms follow.

Sun's Java Accessibility

Sun's Java Accessibility platform consists of a suite of tools, an API, foundation classes, and a "bridge" that ties an application running in its native operating system platform to the Java Accessibility support entities.

Table 11.1 describes these four key parts.

Table 11.1   Java Accessibility Support

Entity	Description
Accessibility API	The Java Accessibility API provides the developers with the ability to embed popular accessible applications, including screen readers and screen magnification software, within the Windows operating system's applications and applets. The API consists of classes and interfaces that provide hooks that link to assistive technologies. You can find a complete description of the API at `http://java.sun.com/products/jfc/swingdoc-api-1.1/javax/accessibility/package-summary.html`.
Accessibility Utilities	The Java Accessibility Utilities provide location and query support for Java applications within the Java Virtual Machine (VM). The utilities include event listeners, trackers, and loading protocols for the VM. You can download the utilities from `http://java.sun.com/products/jfc/#download-access`.
Accessibility Bridge	The Java Accessibility Bridge is the mechanism that links an application running in its operating system environment to accessibility support located within the Java Virtual Machine. You can find more information about the Accessibility Bridge and how it is implemented at `http://java.sun.com/products/jfc/jaccess-1.0/doc/bridge.html`.
Foundation Classes	The Java Foundation Classes (JFC) are uniquely designed as pluggable, look and feel interfaces. These interfaces have been created in a way that not only promotes accessibility within Java development, but also enables you to easily specify a preferred configuration using audio and tactile presentation tools. This promotes direct accessibility by making information available through alternate modalities. You can find a summary of the interfaces at `http://java.sun.com/products/jfc/swingdoc-api-1.1/javax/swing/plaf/package-summary.html`.

III

11

In addition to the Java Accessibility API, Sun created the Java Speech API, a separate set of interfaces protocols that enable application developers to build speech technology directly into their applications and Java applets. You can find more information about the Java Speech API at `http://java.sun.com/products/java-mecia/speech/index.html`.

IBM provides several Java accessibility resources, including a self-voicing development kit that enables automatic speaking of Java applications, developer guidelines, and a quick reference checklist. You can find links to these resources on the following sites:

- The IBM Self-Voicing Kit: `http://www.alphaworks.ibm.com/tech/svk`.
- IBM Guidelines for Writing Accessible Java Applications: `http://www-3.ibm.com/able/accessjava.html`
- IBM Checklist for Accessible Java Applications `http://www-3.ibm.com/sns/accessjava.html#checklist`

Microsoft Corporation is also involved in the advancement of accessibility using Active Accessibility as noted in the next section.

Microsoft Active Accessibility (MSAA)

Microsoft Active Accessibility (MSAA) (`http://www.microsoft.com/enable/msaa/default.htm`), consists of a series of underlying operating system components that improve the way programs and accessibility aids work together by exposing as much information as possible about the Microsoft operating systems constructs (toolbars, menus, text, graphics, and so on) to the accessibility application or device, as well as to the user. Recently both Adobe (Acrobat) and Lotus (Lotus Notes) have implemented MSAA to increase the accessibility of their products to users with disabilities. This is a tremendous leap forward in software development and accessibility and a strong indication that mainstream software vendors see value in developing accessible applications.

For example, it is crucial for screen readers (applications that read the contents of a computer screen to a blind or visually impaired user) to have total knowledge of a screen's content in order to facilitate easier navigation and overall user interface comprehension. To facilitate this, MSAA not only exposes information about the screen elements, but also exposes information about their relationship to one another and other screen elements.

MSAA provides detailed information about operating entities as described in Table 11.2.

Table 11.2   MSAA Support

Type	Description
User Interface Elements	Controls, toolbars, menus, and dialog boxes. Elements that appear as part of the user interface or within a document.
Document Contents	Text, graphics, annotations, comments, and images within a document.
Document Structure	Description of the document organization including paragraphs, lists, sections, and tables.

Through MSAA, and by creating development guidelines for software engineers and application developers, Microsoft is promoting mainstream application accessibility for all technologies. Additionally, accessibility applications and devices are better equipped to deal with new updates and user interface changes at the operating system and application level.

This is especially useful as the current Windows operating system's desktop becomes more and more like a Web site. Internet Explorer 3.0 and 4.01 include MSAA support. MSAA comes installed with Windows 98 and the beta versions of Windows 2000. You can install it on Windows 95 and NT platforms.

To promote accessible application development, Microsoft has developed a Software Development Kit (SDK) for MSAA. You can find information about how to obtain and install the kit at `http://www.microsoft.com/enable/msaa/download.htm`.

Using MSAA, Microsoft includes Active Accessibility for Java, a suite of Java interfaces that assists developers in the creation of accessible applications. You can find guidelines and information for creating accessible applications using MSAAJ at `http://www.microsoft.com/java/resource/access.htm`.

III

11

Cross-Reference

Refer to Chapter 9 for information about creating applications using Active Accessibility for Java.

Achieving Pervasive Accessibility

Much of what we've discussed regarding recent trends in emerging information technologies are strong indicators that industry is supporting the development of architectures, interfaces, applications, and standards that support accessibility for people with disabilities. Where is this all leading to? In my opinion, it's leading to the development of an accessible information technology infrastructure that features direct and pervasive accessibility for all people, particularly people with disabilities. The Internet and World Wide Web are the foundation for this development.

During the 1991 World Congress on Technology conference, the late Digital Equipment Corporation (DEC) visionary David Stone gave a captivating presentation involving pervasive accessible technology for people with disabilities. Stone's theme, "Accessible Technology: A Look Towards the Third Wave," identified the weakness of past and current personal assistive technology interfaces and proposed the development of an advanced accessibility architecture.

The weakness of personal assistive technology, Stone revealed, can be directly attributed to two primary problems:

1. Emerging technology grows at faster rates and more powerfully than personal assistive technology.
2. Personal assistive technology adapts and reacts to emerging technology *after* the technology has reached broad usage.

David then stated:

I submit to you that, in order to give disabled persons full access to technology, we have to ride not only the personal assistive technology wave, but also the systems integration and pervasive accessible technology waves of the future.

In simple terms, we need to make provisions for integrating personal assistive products as front ends or access ports to technology, and we need to make information technology pervasively disability friendly!

According to Stone, achieving accessibility on a pervasive level requires an architecture consisting of four key components:

1. An infrastructure that is capable of running on cross-platform, multivendor systems.

2. A layer of integration services that tie the multivendor systems together and include accessibility services including speech recognition, Access Keys, synthetic voice, and so on.

3. A third level containing accessible applications and data.

4. An accessible framework that acts a personalized interpreter to the individual, regardless of that person's abilities or language skills.

The creation of an accessible information architecture cannot be achieved without the implementation of international accessibility standards. These standards must identify the requirements that information technology and its interface can accommodate. Additionally, it is imperative that they are supported by the IT industry at large.

Next generation information technology human interfaces require a set of accessibility standard interfaces for speech, touch, video, and sound (among others). The accessible information technology infrastructure also requires a set of standard accessibility services that are available on all operating system platforms. Data must be stored and maintained in an accessible format. The end result is a pervasive accessibility technology that serves the user, putting the responsibility of adaptation on the information technology (not the user).

Donald Norman recently highlighted this design regarding human-centered product development by saying, "The goal is a technology that serves the user, where the technology fits the task and the complexity is that of the task, not the tool."

Now, consider this architecture in light of what we previously discussed. A National Technology Grid featuring universal access. Next-generation hardware with universal ports supporting accessible technology. Information appliances and services based on common or standard human interfaces with built-in accessibility features. Cross-platform and open software application architectures such as Java and MSAA that provide direct links to accessibility services. Flexible data description languages such the eXtensible Markup Language (XML) that enable content publishers to describe and deliver data any way you want.

The one entity that still requires development is the external framework that functions as a personalized interpreter to every user, anywhere, at anytime. A framework that knows you and your preferences, yet is flexible enough to adapt to changes based on your operational environment and information requests. Once this is accomplished, we will have created a pervasive accessibility architecture that provides the ultimate infrastructure for every user, regardless of ability or language, ubiquitously.

III

11

Summary

In this chapter you explored the future of emerging technology and accessibility. You learned how research and information engineering is beginning to integrate accessible interface standards that will increase access to people with disabilities. You also learned that emerging technology is headed toward the development of a pervasive accessibility architecture that will accommodate all users, regardless of ability.

References

IBM's Java Accessibility, `http://www-3.ibm.com/sns/accessjava.html`.

The IBM Self-Voicing Kit, `http://www.alphaworks.ibm.com/tech/svk`.

The National Education, Outreach, and Training Program, `http://www.eot.org/`

Norman, Donald A. *The Invisible Computer.* Cambridge: The MIT Press, 1998.

Stone, David. "Accessible Technology: A Look Towards the Third Wave." World Congress on Technology Proceedings, 1991.

The Universal Design/Disability Access for Advanced Computational Infrastructure, `http://trace.wisc.edu/world/udda/`.

WebTV, `http://www.webtv.com/`.

Chapter 12

Web Accessibility Resources

In This Chapter

- WAI-ing the difference
- Identifying industry supporters
- Interesting research organizations
- Accessing government agencies
- Contacting charities and nonprofits
- Disability support organizations
- Other useful sites

The Web is filled with resources that specialize in Web accessibility. In this chapter, you are exposed to organizations that are deeply involved in Internet and WWW access development, education, and research. Several have developed their own Web sites, and many others have developed products and services that can assist you in the development of your own accessible Web site.

Also note that I have not limited this section to resources located in North America. The community of people with disabilities stretches across many boundaries. The significance of this is evident by the sheer number of Web accessibility resources available today. Because there are so many resources, the intent of this chapter is to provide you with a reliable collection of Web accessibility experts, rather than an exhaustive catalogue. Using this list, you will be able to access thousands of other Web sites devoted to accessibility.

All the sites listed can be found on WebABLE! (http://www.webable.com/). I encourage you to surf each site and take advantage of the wealth of information, products, and tools each has to offer. Many of the organizations described here were kind enough to provide their own mission statements and business descriptions.

WAI-ing the Difference

Beyond the use of WebABLE! as a key online resource for Web accessibility (sorry, I couldn't resist the plug), the central resource for Web accessibility development is the Web Accessibility Initiative (http://www.w3.org/WAI/). The WAI's program office, directed by Judy Brewer, has worked overtime to ensure that the WAI's charter and mission is tightly focused and successful.

It is highly advisable that you follow WAI activities if your goal is to develop a Web site accessible to people with disabilities.

The World Wide Web offers the promise of breaking down many traditional barriers to information and interaction among different peoples. The W3C's commitment to lead the Web to its full potential includes promoting a high degree of usability for people with disabilities. The Web Accessibility Initiative, in coordination with other organizations, is pursuing accessibility of the Web through five primary areas of work: technology, guidelines, tools, education and outreach, and research and development.

Identifying the Industry Supporters

After the release of the Mosaic graphical browser, fear that the Web would become inaccessible spread among those within the blind and visually impaired community. At that time, they were just beginning to overcome the complexities of the graphical user interface introduced with the Windows operating environment. Would they have to start all over again?

No. Today, corporations, software vendors, and solution providers are releasing new and enhanced products that assist the Web surfer who has a disability. Recently, several software products including browsers, authoring tools, and plug-ins are appearing with accessibility features. In some cases, whole user interfaces of popular off-the-shelf products are being modified to increase the level of accessibility to the disabilities consumer market.

The following businesses are just a few of the companies supporting Web accessibility within their product lines.

SoftQuad Software

SoftQuad Software's (http://www.softquad.com/) HoTMetaL PRO not only holds the distinction of being the first publicly available Web publishing tool, but it is also the only Web publishing application that inherently supports accessibility. HoTMetaL PRO support includes built-in Web accessibility validation, prompting, and the Visual Dynamic Keyboard (VDK). (Refer to Chapter 8 for additional information about using HoTMetaL PRO and its accessible features.)

IBM Special Needs Systems

IBM Special Needs Systems (http://www-3.ibm.com/able/) has a vision. IBM sees technology as a way to enhance the employability, education, and quality of life of people who have disabilities. Under the Independence Series trademark, IBM has developed a number of assistive devices and software tools that make the computer more accessible and friendly to people who have vision, hearing, speech, mobility, and attention/memory disabilities. These products include everything from interactive speech and cognitive therapy tools to screen reader software, keyboard access utilities, and screen enlargement programs.

IBM's vision is one of all people — teachers, parents, vocational rehabilitation professionals, various associations, advocacy groups, governments, the medical community, and people who have disabilities — working together to use technology to open doors for achievement and independence.

III

12

Microsoft Accessibility and Disabilities

Microsoft's Accessibility and Disabilities group (http://www.microsoft.com/enable/) continues to push ahead in its effort to increase accessibility for all Microsoft products, including Internet Explorer and SAMI (Synchronized Accessible Media Interchange).

To ensure accessibility of its products and to encourage product accessibility by Microsoft partners, Microsoft promotes a four-pronged strategy:

- Make it easier for third-party vendors to create accessibility aids
- Make it easier to make mainstream software accessible
- Promote accessibility through various outreach activities
- Make Microsoft products accessible as mandated by corporate policy

Sun Microsystems' Enabling Technologies Program

Sun's Enabling Technologies program office (http://www.sun.com/access/) is a critical partner in the development of accessible Web interfaces, particularly through its Java application software. The Enabling Technologies office is responsible for developing architectural strategies and solutions to benefit users with disabilities. Sun describes an enabling technology as "one which improves system access, system control, and overall task productivity for ALL users."

The Enabling Technology program's focus is to build an accessibility architecture for Java that provides the following:

- Works with the accessibility architecture(s) built into any platform on which Java programs run
- Scales and extends beyond desktop systems to include Java-powered devices such as information kiosks and nomadic systems
- Enables and promotes built-in accessibility to minimize the need for external assistive technology
- Provides programmatic support for assistive technologies
- Enables developer tools to implement and utilize the architecture in a way that promotes "access friendly" design methods

The BBC's Education Text to Speech Internet Enhancer (Betsie)

While the British Broadcasting Corporation (BBC) is primarily an information broadcasting corporation, the Betsie project (http://www.bbc.co.uk/education/betsie/) has raised an incredible amount of interest about the disabilities community, particularly the blind and visually impaired. Using text-to-speech technology, the BBC is able to modify a Web page and render it using voice output technology. It is a model program for future online news and information agencies to follow.

TIA Access

Technology convergence is just beginning to have an affect on several technologies, including telecommunications networks and products, which are key components to the Web. As a result, the Telecommunications Industry Association (TIA) in collaboration with the Electronic Industries Foundation (EIF) recently developed TIA Access (`http://www.tiaonline.org/access/`) to serve as a clearinghouse for important industry and consumer information on accessible telecommunications technology.

As an organization made up of hundreds of corporations (both large and small), TIA Access is a significant recognition of the importance of Web accessibility.

Interesting Research Organizations

Research and development have played an important role in the design and accessibility of the World Wide Web. From the start, research institutions including NCSA, Trace, ATRC, and CAST have led the charge toward developing a more accessible Web or, to steal a phrase from the Trace Web site, to ensure that we all are "Designing a More Usable World."

Trace Research and Development Center

Trace (`http://trace.wisc.edu/`) is a nonprofit research center that focuses on making off-the-shelf technologies and systems such as computers, the Internet, and information kiosks more accessible for everyone through the process known as universal, or accessible, design. Trace's mission is in great part funded through the National Institute on Disability and Rehabilitation Research (NIDRR), which is part of the U.S. Department of Education. Trace is designated (with core funding from NIDRR) as the Rehabilitation Engineering Research Center (RERC) on Information Technology Access. It is one of 15 such centers scattered throughout the country, each with a different Rehabilitation Engineering focus. Trace is also a partner in the RERC on Telecommunications, along with Gallaudet University and the World Institute on Disability (WID).

The Trace Center has been widely regarded for many years as the leading research, development, and resource center in the area of access to computers by people with disabilities. Over the last several years, the Trace Center has also become well recognized for its work in disability access and universal design of the World Wide Web, information transaction machines, and telecommunications.

III

12

A key focus of the Trace Center's current work is the development of the fundamental underlying technologies needed to ensure access to the emerging information technologies. Collaborative efforts supporting emerging technology projects include work with Sun Microsystems and IBM on the Java language and tools, new telecommunications and software access initiatives with Microsoft, the development of advanced Web technologies with the National Computational Science Alliance and the W3C, and research involving usability and new interface design in collaboration with the National Research Council.

The Adaptive Technology Resource Centre (ATRC)

The Adaptive Technology Resource Centre (ATRC) (http://www.utoronto.ca/atrc/) works directly with information technology manufacturers and developers to influence the early design stages of tomorrow's computer-based technology. The ATRC uses a user-centered design approach to model and create solutions that are commercially feasible, operationally effective, and universally accessible. Areas of research include, but are not limited to the following:

- Access to the Internet and the World Wide Web (browsers, authoring tools, and content)
- Alternative computer display systems
- Alternative computer control methods
- Accessible distance education and videoconferencing
- Equal access to computer-mediated 3D environments
- Gesture recognition for access to communication

NCSA's Mosaic Access Project

The NCSA Mosaic Access Project (http://bucky.aa.uic.edu/) is a resource for those interested in how people with disabilities can use the Internet and the World Wide Web. This page was developed under the Mosaic Access Project with support from the National Science Foundation. The Mosaic Access Project was the first official project to focus on Web accessibility for people with disabilities. The short-term goals of this project were to identify some of the major barriers people with disabilities encountered when using NCSA Mosaic. Where feasible, solutions were designed and implemented.

The long-term goals involved issues of access to the Internet, the World Wide Web, and the accessibility of multimedia browsers in general. Retrofit solutions were developed for Mosaic and barriers inherent to the Web architecture were identified. Recommendations were made to appropriate standards groups and access solutions were explored with current and emerging technologies.

Accessing Government Agencies

The governments of most countries often are responsible for the initial funding of programs, products, and services where the advancement of technology is concerned. The Internet and its succeeding generations (INET 2 and INET 3) are classic examples of government-supported technology.

Accessibility of the Web is no different. In 1997 and 1998, government organizations were predominantly responsible for funding major initiatives involving access to the Web for people with disabilities, including the Web Accessibility Initiative. Additionally, governments in Europe, Asia, Australia, and North America have contributed through legal and standards support. No doubt efforts in Africa and South America will soon be stirred up (based on rumors I've recently heard).

At least three government agencies deserve special mention because they have directly contributed to Web accessibility on two fronts: the U.S. National Institute of Standards and Technology (NIST), the National Aeronautics and Space Administration (NASA), and Canada's Treasury Board Secretariat.

Note

International and national government initiatives contributing to Web accessibility development exist in several nations. The examples that follow reflect just a few of these model efforts.

NIST

The Visualization and Virtual Reality Group at NIST is closely involved in the development of accessibility tools using advanced technology to assist people with disabilities. Sandy Ressler and Sharon Laskowski are two people of special note — Sandy for his work involving accessibility of the Virtual Reality Modeling Language (VRML) (`http://ovrt.nist.gov/projects/VRMLaccess/vrml98/vrmlaccess.htm`) and Sharon for her work in the creation of the WebMetrics Tool Suite. An element of the WebMetrics Tool Suite, the Web Static Analyzer Tool (WebSAT), checks the HTML of a Web page for usability and accessibility. WebSAT is described in more detail in Chapter 6 and is one of the many tools I personally use when reviewing Web sites for accessibility.

NASA

NASA Learning Technologies in partnership with the Johnson Space Center (JSC) Learning Technologies Project developed the ILIAD offline search engine (http://prime.jsc.nasa.gov/iliad.html). ILIAD uses electronic mail or a Web-based form to perform simplified Web searches. The interface is easily accessible on the Web. It uses specialized search agents to perform your search, and then sends you the results by e-mail. ILIAD was originally created as a tool for elementary and secondary school teachers. Not long after its inception, it was determined to be quite useful to blind and visually impaired users who were experiencing difficulties using more graphics-intensive search engines.

If you are interested in using the ILIAD service, you can send e-mail to iliad@prime.jsc.nasa.gov. Type "start iliad" in the subject field and leave the body of the message blank. Instructions for using ILIAD will be e-mailed to you.

Treasury Board Secretariat of Canada

Canada's work involving access to the Web could easily be perceived as taking the lead where government and policy decision makers are concerned. The Treasury Board Secretariat has developed a set of style guidelines, "The Government of Canada's Internet Guide" (http://canada.gc.ca/programs/guide/), that features Web accessibility. This comprehensive set of guidelines is used by Canadian Web site administrators and developers.

The Government of Canada Access Working Group reports to the Treasury Board Secretariat's Internet Advisory Committee regarding all aspects of Web accessibility and assistive technology that supports access to the Web.

At the time of this writing, the Treasury Board Secretariat was developing policy that would require all federal agencies to comply with priority one and two checkpoints of the W3C Web Content Accessibility Guidelines.

Contacting Charities and Nonprofits

Much of the leadership involving Web accessibility development is lead by charities and nonprofit organizations. Over the past couple of years, work involving captioning, validation, and education has been spearheaded by these organizations. Subsequently, the level of Web accessibility awareness has increased immeasurably.

The NCAM, CAST, and the HTML Writers Guild are stellar examples of organizations devoted to ensuring Web access for all people.

CPB/WGBH National Center for Accessible Media (NCAM)

The CPB/WGBH National Center for Accessible Media (NCAM) (`http://www.wgbh.org/wgbh/pages/ncam/`) is a research and development facility dedicated to making media accessible to disabled people in their homes, workplaces, schools, and communities. NCAM and its sister organizations, The Caption Center and Descriptive Video Service (DVS), make up the Media Access division of the WGBH Educational Foundation.

NCAM is also responsible for the development of MAGpie, an important tool used to embed captioning into streaming audio and video. Information about MAGpie can be found on the NCAM homepage at: `http://www.wgbh.org/wgbh/pages/ncam/webaccess/magindex.html`.

NCAM's Web site includes information on projects, publications, and the NCAM Business Partners Program.

The Center for Applied Special Technology (CAST)

Founded in 1984, CAST (`http://www.cast.org/`) is a not-for-profit educational organization whose mission is to expand opportunities for people with disabilities through innovative uses of computer technology. CAST pursues this mission through research, product development, and model educational programs that further universal design for learning.

CAST's Web site reflects their commitment to universal design for learning. They hope to engage others in their ongoing exploration by presenting ideas, illustrative examples, challenging questions, scaffolds, and opportunities for others to contribute. CAST intends that the site be a universally designed, interactive learning environment of the kind they believe should be available to students worldwide. An important element of the Web site is the Bobby home page at `http://www.cast.org/bobby/`. Reflecting CAST's

III

12

commitment to a universally designed Web, Bobby is a free, Web-based program that not only checks Web sites for conformity to international accessibility guidelines for individuals with disabilities, but also provides specific recommendations to correct access problems. CAST is committed to Bobby as a free public service so that it benefits the greatest number of people throughout the world.

HTML Writers Guild

The HTML Writers Guild (http://www.hwg.org/) is the world's largest international organization of Web authors with over 85,000 members in more than 130 nations worldwide. The Guild is devoted to educating their members on all aspects of Web design, including a special commitment to Web accessibility. The HTML Writers Guild teaches online classes and provides Web resources relating to accessible Web design. The Guild's membership dues are reduced by 50 percent for members with disabilities, and free membership is also available.

The HTML Writers Guild exists to assist members in developing and enhancing their abilities as Web authors; to compile and publicize information about standards, practices, techniques, competency, and ethics as applied to Web authoring; and to contribute to the development of the Web and Web technical standards and guidelines.

Disability Support Organizations

Following is a brief list of organizations supporting people with disabilities and focusing on advancement of Web accessibility technology.

Equal Access to Software and Information (EASI)

EASI's (http://www.rit.edu/~easi/) mission is to serve as a resource to the education community by providing information and guidance in the area of access-to-information technologies by individuals with disabilities. EASI stays informed about developments and advancements within the adaptive computer technology field and spreads that information to colleges, universities, K-12 schools, libraries, and industry. EASI is the recipient of two National Science Foundation grants to disseminate information on science, engineering, and math to people with disabilities. EASI presents effective online workshops and hosts frequent live presentations by accessibility experts.

Starling Access Services

Starling Access Services (`http://www.starlingweb.com/`) is a site developed and managed by Web accessibility expert Chuck Letourneau. In addition to providing a wealth of information regarding access to the Web on his site, Chuck takes a novel approach toward teaching Web accessibility — he shows users how to make their sites inaccessible!

Starling Access Services provides seminars and workshops on Web accessibility and is a direct contributor to the WAI.

BrailleNet

BrailleNet is a French Web site providing very useful information about access to the Web for blind Web users. A specific area on their site titled "Better Access to the Web" provides a host of recommendations, resources, and tips for Web accessibility.

Harvey Bingham's Web Accessibility Page

Harvey Bingham is one of those Web accessibility professionals whose expertise spreads across many spectrums. Harvey is an authority on markup languages and is currently involved in developing an XML Document Type Definition for the next generation digital talking book for the Digital Audio Information System (DAISY) Consortium, the Recording for the Blind and Dyslexic, the U.S. Library of Congress National Library Service for the Blind and Physically Handicapped, and the National Information Standards Organization. Harvey also works with the Web Accessibility Initiative of the World Wide Web Consortium.

Harvey's Web page is all text — no bells or whistles — but loaded with useful information. You can find his Web site at `http://www.tiac.net/users/bingham/accessbl/index.htm`.

Other Useful Sites

Following are additional links to important Web accessibility sites that were not mentioned in Part II of this book.

- **The Web Access Gateway**
  `http://www.flatline.org.uk/~silas/access.html`
  This site, created by Silas Brown, functionally intercepts your browser's request to access a Web site and then converts that Web site's page to

III

12

make it more easily readable to screen readers and text browsers. This site also supports internationalization for approximately 30 languages.

- **LIFT**
  http://www.usablenet.com/index.htm
  UsableNet.Com's LIFT is a new web usability and accessibility service. LIFT is available both as an off-the-shelf product or web service. You can access LIFT from WebABLE.

- **ALTifier**
  http://www.vorburger.ch/projects/alt/
  Michael Vorburger's text-only HTML converter, ALTifier, is a tool that generates alternative text for images and other graphical elements on a Web page when those images do not include the ALT attribute.

- **Useit.com**
  http://www.useit.com/
  Jakob Nielsen's Web site is primarily devoted to usability and the Web. Jakob has contributed interesting insights to the Web and is considered the leading expert on Web usability. In my mind, there's a very fine line between usability and accessibility. One begets the other.

- **ZDNet's <devhead> on Accessibility**
  http://www.zdnet.com/devhead/filters/accessibility/
  This is another useful site describing Web accessibility practices and providing links to various Web accessibility resources.

- **Brian Sellden's Linux for Blind Users Web Page**
  http://www.henge.com/~brian/
  Linux is definitely an up-and-coming Web-centered operating system. Brian Sellden's Linux page for blind users provides information about the operating system and accessibility products. Don't be fooled by the name of the Web page listed here and its actual title, "Speech Access for the Masses" — this is the correct address.

- **All Things Web**
  http://www.pantos.org/atw/35263.html
  This is a site devoted to Web design and usability that includes a nice section on accessibility.

- **The Web Design Group**
  http://www.htmlhelp.com/
  This site's address and slogan (". . . Making the Web accessible to all") should tell you that the WDG is a reliable resource. The site is maintained by Liam Quinn, a former key engineer at SoftQuad and contributor to the development of the SGML, HTML, and XML standards. Liam

is very knowledgeable about Web accessibility for people with disabilities.

- **EasyAccess.com**
  `http://www.easyaccess.com/index.html`
  EasyAccess.com is a site devoted to providing Web accessibility services to the blind and low vision Web surfer.

Summary

In this chapter you were introduced to the recognized leaders involved with Web accessibility development. Similar to other major initiatives supporting the Web infrastructure, Web accessibility is a moving target. Be sure you bookmark the sites listed in this chapter to keep up with future developments.

III

12

Chapter 13

Disability Resources

In This Chapter

- Disability organizations
- Charities and nonprofits
- Education and research sites
- Government resources
- Commercial sites
- Informational sites

There is no way that I could have completed this book without the aid of the hundreds of contacts and associates I have developed over the years. Almost all maintain their own Web sites (who doesn't?). The sites listed in this chapter are strictly those that support the disabilities/accessibility field. Please note that this list doesn't begin to touch all of the existing organizations, nationally or internationally. Time and space would not allow me to list all available disability sites — that's a book in and of itself. If you're interested in a particular kind of site, please refer to the WebABLE! disabilities database at `http://www.webable.com` for an exhaustive listing of disability sites. For additional information about WebABLE!, please refer to Appendix A.

As noted in the last chapter, many of the following Web site descriptions were supplied by the individual or organization running the site. I've sorted entries out according to my own definition of where each fits, but in some cases, entries could appear in several categories. Additionally, some sites include descriptions, while others just include the title page and address. No special magic here — I just randomly decided to describe some sites for variety. The first site in the list is always described.

Disability Organizations

Tens of thousands of organizations working in behalf of people with disabilities exist worldwide. As the Web becomes the tour-de-force that drives interaction, collaboration, communication, and information dissemination, disability organizations find it increasingly advantageous to establish a Web presence.

Following is list of key organizations supporting people with disabilities through the Web.

The United Nations Persons with Disabilities Web Page

The United Nations created this Web page (`http://www.un.org/esa/socdev/enable/`) to assist in the promotion of effective measures for the prevention of disability, rehabilitation, the realization of the goals of full participation of disabled persons in social life and social development, and equality.

This is one of the few sites focusing on world programs, standards, and international "norms" for people with disabilities. Important internal links include International Norms and Standards, Global Action, Briefings, and Outreach and Communications.

Deaf World Web

Deaf World Web (DWW, `http://dww.deafworldweb.org/`) is the most comprehensive deaf-related resource on the Internet. Designed as a Web portal, DWW provides an online news service, a database of resources, links to international deaf world Web sites, and an embedded search engine.

One of the more intriguing aspects of this site is its "pop-picks" series of links. I highly recommend spending time in this area, particularly if you are not familiar with the deaf culture. Pop-picks includes the following:

- An animated American Sign Language (ASL) dictionary
- Deaf Culture, an informative series of links relating to the various life style interests of deaf people
- Deaf Kids, a terrific educational resource for kids and teachers
- An e-mail search directory
- A products and services database

The only drawback to this site is that it is designed with HTML FRAMES, making it difficult for some users to access. DWW includes the NOFRAMES alternative, however, which simply provides a text message indicating that users require a frames-compatible Web browser to render the site's information.

Royal Victorian Institute for the Blind (RVIB)

RVIB (http://www.rvib.org.au/) was the first agency in Victoria to work with blind and vision impaired people. Established in 1866, RVIB provides a wide range of services including employment, training, rehabilitation, education, and library and information services.

RVIB's mission is to be the leading provider of services, resources, and information that enable people who are blind or vision impaired to maximize their independence and quality of life.

National Industries for the Blind

The mission of National Industries for the Blind (http://www.nib.org/) is to enhance the opportunities for economic and personal independence of people who are blind, primarily through creating, sustaining, and improving employment.

The Disability Resources Monthly

Disability Resources, Inc., (http://www.disabilityresources.org/) a nonprofit organization established to promote and improve awareness, availability, and accessibility of information for independent living, produces the Disabilities Resource on the Internet. The Web site is designed to help people find the best disability information on the Internet. The main sections are the DRM WebWatcher, a huge subject guide to the best national and

III

13

international disability-related Web sites, and the DRM Regional Resource Directory, an extensive guide to disability organizations and agencies in each state.

American Foundation for the Blind (AFB)

A nonprofit organization founded in 1921 and recognized as Helen Keller's cause in the United States, the American Foundation for the Blind (AFB, http://www.afb.org/) is a leading national resource for people who are blind or visually impaired, the organizations that serve them, and the general public. The mission of the American Foundation for the Blind is to enable people who are blind or visually impaired to achieve equality of access and opportunity that will ensure freedom of choice in their lives.

National Federation of the Blind (NFB)

Founded in 1940, the National Federation of the Blind (NFB, http://www.nfb.org/) is the nation's largest and most influential membership organization of blind people.

The purpose of the National Federation of the Blind is twofold: to help blind people achieve self-confidence and self-respect, and to act as a vehicle for collective self-expression by the blind. By providing public education about blindness; information and referral services; scholarships; literature and publications about blindness; aids, appliances, and other adaptive equipment for the blind; advocacy services and protection of civil rights; job opportunities for the blind; development and evaluation of technology; and support for blind people and their families, members of the NFB strive to educate the public that the blind are normal individuals who can compete on equal terms.

The Low Vision Network (LOVNET)

LOVNET (http://vision.psych.umn.edu/www/lovnet/lovnet.html) is the World Wide Web interface for the Low Vision Research Group (LVRG), providing an Internet resource for researchers, clinicians, and others with an interest in low vision. Here you can learn about recent and ongoing research in low vision, and find links to FAQs, support groups, discussion groups, and other resources for the low vision community.

The Hong Kong Society for the Blind

Established in 1956, the Hong Kong Society for the Blind (http://www.hksb.org.hk/) is a government voluntary agency dedicated to the well being of visually impaired people in Hong Kong. For the past 40 years, the Society has provided comprehensive educational, rehabilitation, vocational training, medical, social, and residential services to visually impaired clients.

The American Council for the Blind (ACB)

The ACB (http://www.acb.org/) "strives to improve the well-being of all blind and visually impaired people by: serving as a representative national organization of blind people; elevating the social, economic and cultural levels of blind people; improving educational and rehabilitation facilities and opportunities; cooperating with the public and private institutions and organizations concerned with blind services; encouraging and assisting all blind people to develop their abilities and conducting a public education program to promote greater understanding of blindness and the capabilities of blind people."

The Rehabilitation Engineering and Assistive Technology Society of America (RESNA)

The Rehabilitation Engineering and Assistive Technology Society of America (RESNA, http://www.resna.org/) is an association of rehabilitation professionals dedicated to the advancement of rehabilitation and assistive technologies for people with disabilities.

The ALS Association

The ALS (amyotrophic lateral sclerosis) Web site (http://www.alsa.org/) is an important effort in the attempt to build awareness about ALS, better known as Lou Gehrig's Disease.

The ALS Association describes their purpose as follows:

The ALS Association seeks to promote awareness and understanding of ALS and the work of The ALS Association by providing up-to-date information and education materials to the ALS community . . . ALS patients and families, caregivers, researchers and members in the health care fields. ALS is a fatal neurodegenerative disease.

III

13

Additional Disability Organizations

- The American Academy of Ophthalmology's EyeNet Low Vision FAQs
  http://www.eyenet.org/public/faqs/low_vision_faq.html
- The Canadian Institute for the Blind (CNIB)
  http://www.cnib.ca/
- DB-LINK, The National Clearinghouse on Children Who Are Deaf-Blind
  http://www.tr.wosc.osshe.edu/dblink/index.htm
- The Royal National Institute for the Blind
  http://www.rnib.org.uk/

Charities and Nonprofits

Similar to organizations dedicated to the support of people with specific types of disabilities, charities and nonprofits abound on the Web. Products, services, and educational outreach make up just a small part of the value-added efforts these organizations provide.

Listed are few of those organizations and a brief description of their offerings.

American Disability Association

The American Disability Association's mission is to provide a quality information network (http://www.adanet.org/) to meet the informational needs of people with diverse disabilities, their care providers, and support professionals; promote disability culture and awareness in the world; and provide forums for creating and refining social policy while working to bring increased quality of life and greater access to freedom to all people regardless of disability.

The Electronic Industries Foundation (EIF)

The EIF (http://www.eia.org/eif/) has long been involved in the support of people with disabilities. While the primary focus of the EIF promoting educational initiatives and programs for children, EIF is directly responsible for funding TIA Access (the Telecommunications Industry Association's disabilities Web site) and the Resource Guide for Accessible Design of Consumer Electronics (http://www.tiaonline.org/access/guide.html). The Resource Guide is one of the most cited guidelines documents available to the disabilities market today. Molly Mannon, EIF Director, is chiefly responsible for the advancement of disabilities and accessibility programs sponsored by the EIF.

Arkenstone

Arkenstone (http://www.arkenstone.org/), led by disabilities guru and advocate Jim Fruchterman, is a leading provider of reading systems for the blind and visually impaired. Arkenstone is well known for their Open Book reading systems and talking GPS personal locators.

Descriptive Video Service

For eight years, Descriptive Video Service (DVS, http://www.wgbh.org/wgbh/access/dvs/) has been "describing" visual images for people who are blind or have low vision. Through an innovative Emmy Award–winning technique, narrated descriptions of visual elements, actions, costumes, gestures, and scene changes are woven into the pauses of a program's or film's soundtrack.

This free, nationwide service, available to 80 percent of American television households, is brought to you by Boston Public Broadcaster WGBH. Since 1990, DVS has described more than 1,600 programs on the Public Broadcasting Service (PBS). For cable customers, DVS has described more than 50 classic movies that appear weekly on the Turner Classic Movies (TCM) cable network.

In addition to PBS programming, more than 175 popular movies and documentaries are available on home video with DVS. These videotapes require only a regular video cassette player and may be borrowed at more than 1,200 libraries nationwide, rented at 500 Blockbuster video store locations, or purchased through the DVS Home Video Catalogue.

Very Special Arts (VSA) of Massachusetts

VSA (http://www.vsamass.org/) is a nonprofit organization that seeks to create and promote opportunities in the arts and cultural mainstream for people with disabilities. VSA also supports the Access Expressed Network (http://www.accessexpressed.net/), a site devoted to disseminating information about a variety of arts and cultural opportunities and the access available to them. The site is designed to contain access information for states across the country.

III

13

NISH

NISH (http://www.nish.org/) enhances the quality of life of people with severe disabilities through increasing employment opportunities. NISH provides professional and technical assistance to not-for-profit community rehabilitation programs to encourage and assist participation in the JWOD program or other appropriate employment or training activities.

Family Village

The Family Village Web site (http://www.familyvillage.wisc.edu/index.htmlx) provides valuable information for parents of individuals who have disabilities. The site describes itself as follows:

> a global community that integrates information, resources, and communication opportunities on the Internet for persons with mental retardation and other disabilities, for their families, and for those that provide them services and support. Our community includes informational resources on specific diagnoses, communication connections, adaptive products and technology, adaptive recreational activities, education, worship, health issues, disability-related media, and literature, and much, much more!

The National Cristina Foundation (NCF)

The National Cristina Foundation (http://www.cristina.org/index2.html) is a nonprofit organization dedicated to providing computer equipment and training to children with disabilities. NCF describes its goals and mission as follows:

> The goal of National Cristina Foundation is to ensure access to computer technology and the sharing of technology solutions to give people with disabilities, students at risk and the economically disadvantaged the opportunity, through training, to lead more productive lives.

Recordings for the Blind and Dyslexic (RFB&D)

RFB&D (http://www.rfbd.org/) is an internationally recognized organization providing professionally narrated and electronic text versions of thousands of books. They maintain a lending library of academic and professional textbooks on audiotape from elementary through post-graduate and professional levels.

One of RFB&D's employees, George Kerscher (http://www.montana.com/kerscher/), has been recognized by the Web community as a visionary in the work of digital recordings for the Web (SMIL). George and

I have often worked together over the past 10 years. Along with Yuri Rubin-sky, we established the International Committee for Accessible Document Design (ICADD) in 1991. ICADD was responsible developing the first accessible electronic standard for markup languages (SGML; ICADD-22).

Currently George serves as the DAISY Consortium Project Director for next generation of digital talking books. George maintains his own Web page at `http://www.montana.com/kerscher/`.

STARBRIGHT Foundation

Few people are aware that Steven Spielberg and General H. Norman Schwarzkopf chair a foundation dedicated to supporting children with special needs. The STARBRIGHT Foundation's Web site (`http://www.starbright.org/`) describes their mission as follows:

> The STARBRIGHT Foundation is dedicated to the development of projects that empower seriously ill children to combat the medical and emotional challenges they face on a daily basis. STARBRIGHT projects do more than educate or entertain: they address the core issues that accompany illness — the pain, fear, loneliness, and depression that can be as damaging as the sickness itself.

Additional Charities and Nonprofits

- The Ability Project
  `http://www.ability.org.uk/home.html`
- The British Computer Association for the Blind
  `http://www.bcab.org.uk/`
- Closing the Gap
  `http://www.closingthegap.com/`
- The Graphic Arts Guild Associations Disability Access Symbols Page
  `http://www.gag.org/resources-das/index.html`
- The Massachusetts Assistive Technology Partnership Program (MATPP)
  `http://www.matp.org/`
- The National Organization on Disability
  `http://www.nod.org/`
- The New England Regional Assistive Technology Exchange
  `http://www.reply.net/cgi-bin/nhaat/index`
- The Upshaw Institute for the Blind
  `http://www.upshawinst.org/`

III

13

Education and Research Sites

Accessibility on an international scale has been greatly influenced by the efforts of education and research institutions pushing the envelope to make accessibility a part of everyone's life. More importantly, many of the organizations listed in this section are responsible for shifting the focus from a person's disability to his or her ability and empowerment in life.

The National Center for Dissemination of Disabilities Research (NCDDR)

The National Center for the Dissemination of Disability Research (NCDDR, http://www.ncddr.org/) is a pilot project designed to help researchers who are funded by the National Institute on Disability and Rehabilitation Research (NIDRR) publicize the results of their research.

The Assistive Technology and Human-Computer Interaction Laboratory at the Institute of Computer Science (ICS) Foundation for Research and Technology — Hellas (FORTH) in Hellas, Greece:

Human-Computer Interaction (HCI) and usability are two of my biggest pet peeves where interface design for people with disabilities is concerned. Very little has been accomplished in the area of information and advanced technology with regard to HCI research and people with disabilities. The AT-HCI Lab at ICS-FORTH is a recognized leader in research and development of HCI for the disabled (http://www.ics.forth.gr/proj/at-hci/). The Laboratory describes its mission as follows:

> The overall objective of R&D activities at the AT&HCI Lab, ICS-FORTH, is to contribute towards the development of an Information Society acceptable to all citizens. Research and development work at the AT & HCI Lab in the fields of Assistive Technology and Human-Computer Interaction aim to ensure that the user interfaces of the emerging telematic applications and services are accessible and usable by all potential users, including disabled and elderly people.

> Based on the principles of Universal Accessibility and Design for All, current research efforts at the AT & HCI Lab focus on the development of User Interfaces for All (a concept recently proposed in the international literature by the Lab), which provide accessibility and high quality of interaction to the user population at large, including disabled and elderly people. In this context, the Lab has provided the technical framework for the development of Unified User Interfaces which are adapted to the end-user abilities, skills, requirements and preferences.

The DAISY Consortium

The Digital Audio-based Information System (DAISY) Consortium is establishing the International Standard for the production, exchange, and use of the next generation of digital talking books. The DAISY Consortium consists of organizations worldwide serving people who are blind or print disabled. (People with print disabilities are individuals who cannot easily read printed material. For example, a person who cannot turn pages of a book or an individual with dyslexia.)

Disability Information Systems in Higher Education (DISinHE)

This site (http://www.disinhe.ac.uk/) is the Central Clearing House for Information Technology for Accessibility and Disability in Higher Education, funded by the Joint Information Systems Committee (JISC) of the United Kingdom Higher Education Funding Councils.

Gallaudet University's Technology Access Program

This is a terrific site (http://tap.gallaudet.edu/) featuring research, services, and technology involving communications, telecommunications, and the deaf.

Special Needs Opportunity Windows (SNOW)

The SNOW Web site (http://snow.utoronto.ca/) is a Canadian project funded to provide an information clearinghouse for educators of students with special needs. In addition to containing a wealth of resource information, SNOW provides discussion forums and a kid's area to assist children with disabilities.

SNOW has hosted WebABLE! for the past two years and has been a valuable aid in the promotion of awareness and education of Web accessibility.

Center for Research on Women with Disabilities

The purpose of the Center for Research on Women with Disabilities (http://www.bcm.tmc.edu/crowd/crowd1.html) is to conduct research and promote, develop, and disseminate information to expand the life choices of women with disabilities so they may fully participate in community life.

III

13

The Institute on Disability at University of New Hampshire

This organization (http://www.iod.unh.edu/), which was established in 1987, is an important resource for disability research and information. The Institute on Disability provides a host of services including in-house training for students and professionals, technical assistance programs, assistive technology research, and general information awareness activities aimed at educating the public about disability topics.

The University of Illinois Urbana-Champaign Information Technology Accessibility Page

This is a great university and educational resource for accessibility research (http://www.als.uiuc.edu/InfoTechAccess/), particularly in the areas of advanced computing, information services, and the Web.

Additional Education and Research Sites

- The LaGrange Area Department of Special Education
  http://www.ladse.k12.il.us
- Riverdale High School Web Accessibility Page
  http://members.home.com/davidjarvis/
- The Society for Disability Studies
  http://www.uic.edu/orgs/sds/
- The Universal Design/Disability Access Program for Advanced Computational Infrastructure
  http://trace.wisc.edu/world/udda/

U.S. Government Resources

Some people argue that accessibility is only effective when government creates legislation and standards that enforce access for people with disabilities. Conversely, others argue that government enforcement impedes ingenuity and entrepreneurial efforts. Regardless of your personal view, several government agencies have funded efforts to increase Web accessibility awareness.

The following sections include descriptions of several U.S. and international government organizations with established Web sites.

National Institute on Disability and Rehabilitation Research (NIDRR)

NIDRR (http://www.ed.gov/offices/OSERS/NIDRR/index.html) sits on the top of my list of government resources not because they are an important resource, but rather because they fund hundreds of programs that ultimately become valuable resources to people with disabilities. Very few U.S. disability programs exist without some sort of funding from NIDRR. An agency within the Department of Education, NIDRR is directly responsible for funding the Web Accessibility Initiative, Trace Research, and the National Center for Dissemination of Disability Research (NCDRR).

General Services Administration's Center for IT Accommodation (CITA)

GSA's Center for IT Accommodation (CITA, http://www.itpolicy.gsa.gov/cita/) is a nationally recognized model demonstration facility influencing accessible information environments, services, and management practices.

The President's Committee on Employment for People with Disabilities

The President's Committee (http://www50.pcepd.gov/pcepd/) is a small federal agency whose Chairman and Vice Chairs are appointed by the President of the United States. The mission of the committee is the following:

> to facilitate the communication, coordination and promotion of public and private efforts to enhance the employment of people with disabilities. The Committee provides information, training, and technical assistance to America's business leaders, organized labor, rehabilitation and service providers, advocacy organizations, families and individuals with disabilities.

The National Council on Disability (NCD)

The National Council on Disability (http://www.ncd.gov/) is an independent U.S. federal agency that makes recommendations to the President of the United States and the U.S. Congress on issues affecting Americans with disabilities.

NCD's mission is as follows:

> NCD's overall purpose is to promote policies, programs, practices, and procedures that guarantee equal opportunity for all individuals with disabilities, regardless of the

III

13

nature of severity of the disability; and to empower individuals with disabilities to achieve economic self-sufficiency, independent living, and inclusion and integration into all aspects of society.

Additional U.S. Government Resources

- City of San Jose, California's Access Instructions Web Page
  http://www.ci.san-jose.ca.us/access.html
- Department of Justice ADA Page
  http://www.usdoj.gov/crt/ada/
- Maryland Division of Rehabilitation Services
  http://www.dors.state.md.us
- NIDRR's ABLEDATA
  http://www.abledata.com/

International Government Resources

Following is a brief list of international government resources. Refer to Chapters 2 and 3 for additional international resources.

The Government of Canada Internet Guide

Canada's Treasury Board Secretariat has published a comprehensive set of Internet and Web design guidelines (http://www.gc.ca/programs/guide/main_e.html) for government Web site managers and developers.

Disability Information and Resource Center for South Australia (DIRC)

DIRC (http://gateway.dircsa.org.au/) is one of the leading government-run disabilities information sites available on the Web today. This site epitomizes a fascinating aspect of the Web: international barriers seem to tumble when it comes to providing people-specific services. The fact that this site resides in South Australia has bearing on the importance of its information to people with disabilities everywhere. DIRC describes its mission as follows:

> The Disability Information & Resource Centre is an Incorporated body funded by the State Government. The Management Committee includes people with disabilities and representatives of organisations providing services to people with disabilities.

DIRC's role is to "point people in the right direction" — to end the confusion concerning where to go for information about disability — to refer people to the most appropriate place to meet their needs.

DIRC has a clear policy — they are available to all people with disabilities, and to organisations and persons involved in providing services to people with disabilities.

Commercial Sites

For every disability, there's a product or service. The following list of vendors, manufacturers, and service and commercial trade associations will assist you in selecting among the thousands that exist on the Web.

Apple Computer Disability Solutions Group

Apple Computer has long been an advocate and integrator of accessibility products for the desktop computing platform practically from its inception. All Macintosh systems come with built-in accessibility features including CloseView and EasyAccess. On a user-interface level, few systems compare with the easy of use presented through the Macintosh desktop or notebook computers.

In support of their work involving people with disabilities, Apple's Disability Solutions Group is responsible for ensuring that Apple products are accessible to individuals with disabilities (`http://www.apple.com/education/k12/disability/message.html`).

The Assistive Devices Industry Association (CanADIA) of Canada

The Assistive Devices Industry Association of Canada (CanADIA, `http://www.starlingweb.com/canadia/`) is a trade association of assistive technology manufacturers and developers of Canada. CanADIA supports this industry by promoting their interests on a national and international basis.

While the CanADIA organization is an industry-supported organization, their Web site also provides an incredibly useful calendar service that includes a comprehensive listing of disability-related conferences.

Disability Net

Disability Net (`http://www.disabilitynet.co.uk/`) is one of the few Internet service providers that directs its services toward people with disabilities.

III

13

Disability Net also provides a variety of disability-related information services including news, product sales, discussion forums, a job center, and research data.

The vOICe

The vOICe (http://ourworld.compuserve.com/homepages/Peter_Meijer/) is an experimental system for auditory image representations. It was designed as a step toward a vision substitution device for the blind through the real-time conversion of arbitrary images into soundscapes. Note that to use this system, a videocam is required.

Adaptive Technology Consulting, Inc.

Adaptive Technology Consulting (ATC, http://www.adaptivetech.net/) carries a wide selection of products for the blind, visually impaired, and individuals with reading difficulties. ATC provides complete services for system integration as well as training and support on the total system.

Immersion Corporation: http://www.immerse.com/

Immersion Corporation's work on force-feedback products is cutting-edge, to say the least. Force-feedback renders the sensation of "feel" to a mouse, adding a new dimension to the personal computing interface. For example, a blind person could feel his way around a windows desktop as the mouse generates feedback based on objects it touches. Check this page out and review their work with force-feedback devices: http://www.immersion.com/. Their work is truly revolutionary and could be part of the next generation of input device products.

Duxbury Systems, Inc.

Some corporations stand out as legends in any industry. Duxbury Systems, Inc. (http://www.duxburysystems.com/) is one such legendary corporation in the field of assistive technology for the blind. Joe Sullivan, President of Duxbury Systems, is considered an icon in the Braille translation software business, perhaps second only to Louis Braille. The Duxbury Braille translation system is easily one of the most popular in the industry. It is capable of translating Braille into "English, Spanish, French, Arabic, Hebrew, Portuguese, and other languages on MS-DOS, Windows, Macintosh, Unix and other systems."

Techno-Vision Systems Ltd.

Techno-Vision Systems Ltd. (http://www.techno-vision.co.uk/) is a UK-based company providing products and services for visually impaired people, including products for describing tactile audio and graphics.

Enabling Technologies Company (ETC)

ETC (http://www.brailler.com/) provides Braille production equipment that they design, manufacture, service, and support for Braille tasks of every size and variety.

Formation et Insertion Pour Deficients Visuels (FIDEV)

FIDEV's mission is to improve the professional integration of blind and visually impaired people by providing training and integration services (http://fidev.ec-lyon.fr/pages/indxang.html.en).

Lunney Associates

Lunney Associates (http://people.delphi.com/LUNNEY) develops high-technology laboratory aids and data analysis tools for science students, scientists, and engineers who have disabilities, especially those who are blind or visually impaired. They also offer technical consulting on workplace accommodations under the Americans with Disabilities Act (ADA) for scientists and engineers who have disabilities. The site has a more general emphasis on using computer-based tools to make science, engineering, and mathematics more accessible people with disabilities, especially K–16 students.

ActiveWord Systems, Inc.

ActiveWord Systems (http://www.activewords.com/) has taken an interesting approach to computing by making common or familiar words active in all user contexts as you work on your computer. It's an extremely interesting feature that is quite useful to people with disabilities.

Additional Commercial Sites

- Academic Software, Inc.
  http://www.acsw.com/
- Dragon Systems
  http://www.dragonsys.com/

III

13

- Freedom of Speech, Inc.
  http://store.yahoo.com/fos/
- LS&S Group
  http://www.lssgroup.com/
- Synapse
  http://www.synapseadaptive.com/
- Universal Design Institute for Information Technology
  http://www.udit-jp.com
- Yahoo's Society and Culture for Disabilities
  http://www.yahoo.com/Society_and_Culture/Disabilities/
  Organizations/

Sites for the Deaf and Hearing Impaired

- CommWare Technologies
  http://www.commware-tech.ccm/
- Hearing Aid Centers (HAC) of America Group
  http://www.harcmercantile.com/hac01.htm
- MultiMedia Design's Personal Captioning System
  http://www.multimediadesigrs.com/pc.html
- Nationwide Flashing Signal Systems
  http://www.nfss.com/
- NXi Communications, Inc.
  http://www.nxicom.com/

Sites for the Blind and Visually Impaired

- AccessAbility, Inc.
  http://www.customeyes.com/
- Ai Squared
  http://www.aisquared.com/
- Artic Technologies
  http://www.artictech.com/
- Blazie Engineering
  http://blazie.com/
- GW-Micro
  http://www.gwmicro.com/
- Innoventions
  http://www.magnicam.com/magnicam/

- Syntha-Voice
  `http://www.synthavoice.on.ca/`
- TeleSensory
  `http://www.telesensory.com/`
- VisuAide Products
  `http://www.visuaide.com/indexproduits.en.html`

Sites for Mobility and Physically Challenged

- Access Unlimited
  `http://www.accessunlimited.com/`
- Handicapped, Inc.
  `http://www.handicapsinc.com/`
- In Touch Systems
  `http://www.magicwandkeyboard.com/index.html`
- New Mobility Interactive Café
  `http://www.newmobility.com/`

Informational Sites

Some sites exist for the purpose of disseminating information. These are the true educators of accessibility. There's not a single site listed in the following sections that I don't have bookmarked within my Web browser.

The National Rehabilitation Information Center (NARIC)

NARIC (`http://www.naric.com/naric/index.html`) is a valuable information resource that has collected and disseminated the results of federally funded research projects. NARIC's literature collection, which includes commercially published books, journal articles, and audiovisuals, averages around 200 new documents per month. You can find NARIC's instant disability information center at `http://www.naric.com/naric/search/`.

ADA Technical Assistance Program

An extremely valuable site (`http://www.adata.org/`) for corporations who employ people with disabilities, the ADA Technical Assistance Program is a government funded program that "provides information, training, and technical assistance to businesses and agencies with duties and responsibilities under the ADA and to people with disabilities with rights under the ADA."

III

13

Duncan Kinder's Americans with Disabilities Document Center

There aren't too many Web sites that can be credited to just one person, but Duncan Kinder has clearly developed one of the most informative disabilities Web sites available (http://janweb.icdi.wvu.edu/kinder/). The ADA Document Center is a repository for U.S. legal documents regarding disability regulations and standards.

The Disability Statistics Center (DSC)

Located at the University of California, San Francisco, the DSC (http://dsc.ucsf.edu/) provides a plethora of disability statistical information. The center's purpose is to function as ". . . a national center of research and training in disability statistics. The Center has ongoing research projects on the cost of disability, employment and earnings, access to health and long-term care services, housing, mortality, and national statistical indicators on the status of people with disabilities in America."

The Able Channel

The mission of the Able Channel (http://www.tvontheweb.com/channels/able/index.html) is to provide and share information on a variety of subjects concerning people with disabilities. By providing the information in an accessible format, the Able Channel assists in empowering this growing segment of the world population to make value-added choices. The Able Channel's vision is to become the central repository on the Web for storing data about activities, assistive technologies, and general information concerned about and for all types of disabilities. However, the information presented will be in an accessible video format, including audio descriptions and closed captioning. The use of the WWW as a delivery mechanism for this type of product has two main strengths: worldwide coverage and affordability. Both the TV on the Web Network and the Able Channel have the ability to provide this leading edge technology via the Real Network's G2 video player. With the recent implementation of the Real Network's G2 server software, the capability of delivering accessible videos through the Web via video streaming techniques can be realized by all people.

Jeffrey Pledger, President of the Able Channel, is doing a fabulous job with this site, creating an accessible Web platform that delivers on-the-spot news, technology updates, and human interest stories.

The Screen Magnifiers Homepage

The Screen Magnifiers Homepage is where you can find information about screen magnification, as well as download demos, shareware, and freeware.

Tell Us Your Story Discussion Forum

Forum for personal experience stories about disability awareness, rights, and inspiration (http://www.tell-us-your-story.com/).

Deborah Quilter's RSIHelp.com Site

Repetitive Strain Syndrome (RSI, http://www.rsihelp.com/) is one of the fastest growing disabilities in the world, particularly because of the use of computers. The RSIHelp.com site is a terrific resource for people who experience this disability.

The World Health Organization Statistical Information System (WHOSIS)

WHOSIS (http://www.who.int/whosis/) is the World Health Organization's health and health-related statistical information site. This site is a terrific resource that provides disability statistics for nations all over the world.

Additional Informational Sites

- The Adaptive Device Locator System (ADLS)
  http://www.acsw.com/adlsweb1.html
- Talking Newspaper Association, UK
  http://www.tnauk.org.uk/

Summary

In this chapter, you were introduced to a wide range of resources supporting various efforts in the field of disabilities. Web sites for charities, research, education, industry, and government resources are available on an international scale and ready to assist you the moment you visit the site.

If you are interested in listing your site on WebABLE!, send e-mail to webinfo@webable.com.

III

13

Appendix A

The WAI Content Authoring Guidelines

W3C Recommendation 5-May-1999

To download these guidelines or check for the latest version, see `http://www.w3.org/TR/WAI-WEBCONTENT`.

Editors:

Wendy Chisholm, Trace R & D Center, University of Wisconsin—Madison
Gregg Vanderheiden, Trace R & D Center, University of Wisconsin—Madison
Ian Jacobs, W3C

Abstract

These guidelines explain how to make Web content accessible to people with disabilities. The guidelines are intended for all Web content developers (page authors and site designers) and for developers of authoring tools. The primary goal of these guidelines is to promote accessibility. However, following them will also make Web content more available to *all* users, whatever user agent they are using (e.g., desktop browser, voice browser, mobile phone, automobile-based personal computer, etc.) or constraints they may be operating under (e.g., noisy surroundings, under- or over-illuminated rooms, in a hands-free environment, etc.). Following these guidelines will also help people find information on the Web more quickly. These guidelines do not discourage content developers from using images, video, etc., but rather explain how to make multimedia content more accessible to a wide audience.

This is a reference document for accessibility principles and design ideas. Some of the strategies discussed in this document address certain Web internationalization and mobile access concerns. However, this document focuses on accessibility and does not fully address the related concerns of other W3C Activities. Please consult the W3C Mobile Access Activity home page and the W3C Internationalization Activity home page for more information.

This document is meant to be stable and therefore does not provide specific information about browser support for different technologies as that information changes rapidly. Instead, the Web Accessibility Initiative (WAI) Web site provides such information (refer to [WAI-UA-SUPPORT]).

This document includes an appendix that organizes all of the checkpoints by topic and priority. The topics identified in the appendix include images, multimedia, tables, frames, forms, and scripts. The appendix is available as either a tabular summary of checkpoints or as a simple list of checkpoints.

A separate document, entitled "Techniques for Web Content Accessibility Guidelines 1.0" ([TECHNIQUES]), explains how to implement the checkpoints defined in the current document. The Techniques Document discusses each checkpoint in more detail and provides examples using the Hypertext Markup Language (HTML), Cascading Style Sheets (CSS), Synchronized Multimedia Integration Language (SMIL), and the Mathematical Markup Language (MathML). The Techniques Document also includes techniques for document validation and testing, and an index of HTML elements and attributes (and which techniques use them). The Techniques Document has been designed to track changes in technology and is expected to be updated more frequently than the current document.

Note Not all browsers or multimedia tools may support the features described in the guidelines. In particular, new features of HTML 4.0 or CSS 1 or CSS 2 may not be supported.

"Web Content Accessibility Guidelines 1.0" is part of a series of accessibility guidelines published by the Web Accessibility Initiative. The series also includes User Agent Accessibility Guidelines ([WAI-USERAGENT]) and Authoring Tool Accessibility Guidelines ([WAI-AUTOOLS]).

Status of this document

This document has been reviewed by W3C Members and other interested parties and has been endorsed by the Director as a W3C Recommendation. It is a stable document and may be used as reference material or cited as a normative reference from another documents. W3C's role in making the Recommendation is to draw attention to the specification and to promote its widespread deployment. This enhances the functionality and universality of the Web.

The English version of this specification is the only normative version. However, for translations in other languages see `http://www.w3.org/WAI/GL/WAI-WEBCONTENT-TRANSLATIONS`.

The list of known errors in this document is available at `http://www.w3.org/WAI/GL/WAI-WEBCONTENT-ERRATA`. Please report errors in this document to `wai-wcag-editor@w3.org`.

A list of current W3C Recommendations and other technical documents can be found at `http://www.w3.org/TR`.

This document has been produced as part of the W3C Web Accessibility Initiative. The goal of the Web Content Guidelines Working Group is discussed in the Working Group charter.

Table of Contents

    The appendix list of checkpoints is available as either a tabular summary of checkpoints or as a simple list of checkpoints.

1. Introduction

For those unfamiliar with accessibility issues pertaining to Web page design, consider that many users may be operating in contexts very different from your own.

- They may not be able to see, hear, move, or may not be able to process some types of information easily or at all.
- They may have difficulty reading or comprehending text.
- They may not have or be able to use a keyboard or mouse.
- They may have a text-only screen, a small screen, or a slow Internet connection.
- They may not speak or understand fluently the language in which the document is written.
- They may be in a situation where their eyes, ears, or hands are busy or interfered with (e.g., driving to work, working in a loud environment, etc.).
- They may have an early version of a browser, a different browser entirely, a voice browser, or a different operating system.

Content developers must consider these different situations during page design. While there are several situations to consider, each accessible design choice generally benefits several disability groups at once and the Web community as a whole. For example, by using style sheets to control font styles and eliminating the FONT element, HTML authors will have more control over their pages, make those pages more accessible to people with low vision, and by sharing the style sheets, will often shorten page download times for all users.

The guidelines discuss accessibility issues and provide accessible design solutions. They address typical scenarios (similar to the font style example) that may pose problems for users with certain disabilities. For example, the first guideline explains how content developers can make images accessible. Some users may not be able to see images, others may use text-based browsers that do not support images, while others may have turned off support for images (e.g., due to a slow Internet connection). The guidelines do not suggest avoiding images as a way to improve accessibility. Instead, they explain that providing a text equivalent of the image will make it accessible.

How does a text equivalent make the image accessible? Both words in "text equivalent" are important.

- Text content can be presented to the user as synthesized speech, braille, and visually-displayed text. Each of these three mechanisms uses a different sense—ears for synthesized speech, tactile for braille, and eyes for visually-displayed text—making the information accessible to groups representing a variety of sensory and other disabilities.

- In order to be useful, the text must convey the same function or purpose as the image. For example, consider a text equivalent for a photographic image of the Earth as seen from outer space. If the purpose of the image is mostly that of decoration, then the text "Photograph of the Earth as seen from outer space" might fulfill the necessary function. If the purpose of the photograph is to illustrate specific information about world geography, then the text equivalent should convey that information. If the photograph has been designed to tell the user to select the image (e.g., by clicking on it) for information about the earth, equivalent text would be "Information about the Earth". Thus, if the text conveys the same function or purpose for the user with a disability as the image does for other users, then it can be considered a text equivalent.

Note that, in addition to benefitting users with disabilities, text equivalents can help all users find pages more quickly, since search robots can use the text when indexing the pages.

While Web content developers must provide text equivalents for images and other multimedia content, it is the responsibility of user agents (e.g., browsers and assistive technologies such as screen readers, braille displays, etc.) to present the information to the user.

Non-text equivalents of text (e.g., icons, pre-recorded speech, or a video of a person translating the text into sign language) can make documents accessible to people who may have difficulty accessing written text, including many individuals with cognitive disabilities, learning disabilities, and deafness. Non-text equivalents of text can also be helpful to non-readers. An auditory description is an example of a non-text equivalent of visual information. An auditory description of a multimedia presentation's visual track benefits people who cannot see the visual information.

2. Themes of Accessible Design

The guidelines address two general themes: ensuring graceful transformation, and making content understandable and navigable.

2.1 Ensuring Graceful Transformation

By following these guidelines, content developers can create pages that transform gracefully. Pages that transform gracefully remain accessible despite any of the constraints described in the introduction, including physical, sensory, and cognitive disabilities, work constraints, and technological barriers. Here are some keys to designing pages that transform gracefully.

- Separate structure from presentation (refer to the difference between content, structure, and presentation).
- Provide text (including text equivalents). Text can be rendered in ways that are available to almost all browsing devices and accessible to almost all users.
- Create documents that work even if the user cannot see and/or hear. Provide information that serves the same purpose or function as audio or video in ways suited to alternate sensory channels as well. This does not mean creating a prerecorded audio version of an entire site to make it accessible to users who are blind. Users who are blind can use screen reader technology to render all text information in a page.
- Create documents that do not rely on one type of hardware. Pages should be usable by people without mice, with small screens, low resolution screens, black and white screens, no screens, with only voice or text output, etc.

The theme of graceful transformation is addressed primarily by guidelines 1 to 11.

2.2 Making Content Understandable and Navigable

Content developers should make content understandable and navigable. This includes not only making the language clear and simple, but also providing understandable mechanisms for navigating within and between pages. Providing navigation tools and orientation information in pages will maximize accessibility and usability. Not all users can make use of visual clues such as image maps, proportional scroll bars, side-by-side frames, or graphics that guide sighted users of graphical desktop browsers. Users also lose contextual information when they can only view a portion of a page, either because they are accessing the page one word at a time (speech synthesis or braille display), or one section at a time (small display, or a magnified display). Without orientation information, users may not be able to understand very large tables, lists, menus, etc.

The theme of making content understandable and navigable is addressed primarily in guidelines 12 to 14.

3. How the Guidelines are Organized

This document includes fourteen guidelines, or general principles of accessible design. Each guideline includes the following.

- The guideline number.

- The statement of the guideline.
- The rationale behind the guideline and some groups of users who benefit from it.
- A list of checkpoint definitions.

The checkpoint definitions in each guideline explain how the guideline applies in typical content development scenarios. Each checkpoint definition includes the following.

- The checkpoint number.
- The statement of the checkpoint.
- The priority of the checkpoint.
- Optional informative notes, clarifying examples, and cross references to related guidelines or checkpoints.
- A reference to the Techniques Document ([TECHNIQUES]) where implementations and examples of the checkpoint are discussed.

Each checkpoint is intended to be specific enough so that someone reviewing a page or site may verify that the checkpoint has been satisfied.

3.1 Document conventions

The following editorial conventions are used throughout this document.

- Element names are in uppercase letters.
- Attribute names are quoted in lowercase letters.
- References to definitions are in italics.

4. Priorities

Each checkpoint has a priority level assigned by the Working Group based on the checkpoint's impact on accessibility.

[Priority 1]   A Web content developer **must** satisfy this checkpoint. Otherwise, one or more groups will find it impossible to access information in the document. Satisfying this checkpoint is a basic requirement for some groups to be able to use Web documents.

[Priority 2]   A Web content developer **should** satisfy this checkpoint. Otherwise, one or more groups will find it difficult to access information in the document. Satisfying this checkpoint will remove significant barriers to accessing Web documents.

[Priority 3]   A Web content developer **may** address this checkpoint. Otherwise, one or more groups will find it somewhat difficult to access information in the document. Satisfying this checkpoint will improve access to Web documents.

Some checkpoints specify a priority level that may change under certain (indicated) conditions.

5. Conformance

This section defines three levels of conformance to this document:

Conformance Level "A"   All Priority 1 checkpoints are satisfied.

Conformance Level "Double-A"   All Priority 1 and 2 checkpoints are satisfied.

Conformance Level "Triple-A"   All Priority 1, 2, and 3 checkpoints are satisfied.

Note   Conformance levels are spelled out in text so they may be understood when rendered to speech.

Claims of conformance to this document must use one of the following two forms.

Form 1

Specify the following information.

- The guidelines title: "Web Content Accessibility Guidelines 1.0"
- The guidelines URL:
  `http://www.w3.org/TR/1999/WAI-WEBCONTENT-19990505`
- The conformance level satisfied: "A", "Double-A", or "Triple-A".
- The scope covered by the claim (e.g., page, site, or defined portion of a site.).

  Example of Form 1:

  This page conforms to W3C's "Web Content Accessibility Guidelines 1.0", available at `http://www.w3.org/TR/1999/WAI-WEBCON-TENT-19990505`, level Double-A.

Form 2

Include, on each page claiming conformance, one of three icons provided by W3C and link the icon to the appropriate W3C explanation of the claim.

Information about the icons and how to insert them in pages is available at [WCAG-ICONS].

6. Web Content Accessibility Guidelines

Guideline 1. Provide equivalent alternatives to auditory and visual content.

Provide content that, when presented to the user, conveys essentially the same function or purpose as auditory or visual content.

Although some people cannot use images, movies, sounds, applets, etc. directly, they may still use pages that include *equivalent* information to the visual or auditory content. The equivalent information must serve the same purpose as the visual or auditory content. Thus, a text equivalent for an image of an upward arrow that links to a table of contents could be "Go to table of contents". In some cases, an equivalent should also describe the appearance of visual content (e.g., for complex charts, billboards, or diagrams) or the sound of auditory content (e.g., for audio samples used in education).

This guideline emphasizes the importance of providing *text equivalents* of non-text content (images, pre-recorded audio, video). The power of text equivalents lies in their capacity to be rendered in ways that are accessible to people from various disability groups using a variety of technologies. Text can be readily output to speech synthesizers and *braille displays*, and can be presented visually (in a variety of sizes) on computer displays and paper. Synthesized speech is critical for individuals who are blind and for many people with the reading difficulties that often accompany cognitive disabilities, learning disabilities, and deafness. Braille is essential for individuals who are both deaf and blind, as well as many individuals whose only sensory disability is blindness. Text displayed visually benefits users who are deaf as well as the majority of Web users.

Providing non-text equivalents (e.g., pictures, videos, and pre-recorded audio) of text is also beneficial to some users, especially nonreaders or people who have difficulty reading. In movies or visual presentations, visual action such as body language or other visual cues may not be accompanied by enough audio information to convey the same information. Unless verbal descriptions of this visual information are provided, people who cannot see (or look at) the visual content will not be able to perceive it.

Checkpoints

1.1 Provide a text equivalent for every non-text element (e.g., via "alt", "longdesc", or in element content). *This includes:* images, graphical representations of text (including symbols), image map regions, animations (e.g., animated GIFs), applets and programmatic objects, ascii art, frames, scripts, images used as list bullets, spacers, graphical buttons, sounds (played with or without user interaction), stand-alone audio files, audio tracks of video, and video. [**Priority 1**]

For example, in HTML:

- Use "alt" for the IMG, INPUT, and APPLET elements, or provide a text equivalent in the content of the OBJECT and APPLET elements.

- For complex content (e.g., a chart) where the "alt" text does not provide a complete text equivalent, provide an additional description using, for example, "longdesc" with IMG or FRAME, a link inside an OBJECT element, or a description link.

- For image maps, either use the "alt" attribute with AREA, or use the MAP element with A elements (and other text) as content.

Refer also to checkpoint 9.1 and checkpoint 13.10.

1.2 Provide redundant text links for each active region of a server-side image map. [**Priority 1**]

Refer also to checkpoint 1.5 and checkpoint 9.1.

1.3 Until user agents can automatically read aloud the text equivalent of a visual track, provide an auditory description of the important information of the visual track of a multimedia presentation. [**Priority 1**]

Synchronize the *auditory description* with the audio track as per checkpoint 1.4. Refer to checkpoint 1.1 for information about textual equivalents for visual information.

1.4 For any time-based multimedia presentation (e.g., a movie or animation), synchronize equivalent alternatives (e.g., captions or auditory descriptions of the visual track) with the presentation. [**Priority 1**]

1.5 Until user agents render text equivalents for client-side image map links, provide redundant text links for each active region of a client-side image map. [**Priority 3**]

Refer also to checkpoint 1.2 and checkpoint 9.1.

Guideline 2. Don't rely on color alone.

Ensure that text and graphics are understandable when viewed without color.

If color alone is used to convey information, people who cannot differentiate between certain colors and users with devices that have non-color or non-visual displays will not receive the information. When foreground and background colors are too close to the same hue, they may not provide sufficient contrast when viewed using monochrome displays or by people with different types of color deficits.

Checkpoints

2.1 Ensure that all information conveyed with color is also available without color, for example from context or markup. [**Priority 1**]

2.2 Ensure that foreground and background color combinations provide sufficient contrast when viewed by someone having color deficits or when viewed on a black and white screen. [**Priority 2 for images, Priority 3 for text**]

Guideline 3. Use markup and style sheets and do so properly.

Mark up documents with the proper structural elements. Control presentation with style sheets rather than with presentation elements and attributes.

Using markup improperly—not according to specification—hinders accessibility. Misusing markup for a presentation effect (e.g., using a table for layout or a header to change the font size) makes it difficult for users with specialized software to understand the organization of the page or to navigate through it. Furthermore, using presentation markup rather than structural markup to convey structure (e.g., constructing what looks like a table of data with an HTML PRE element) makes it difficult to render a page intelligibly to other devices (refer to the description of *difference between content, structure, and presentation*).

Content developers may be tempted to use (or misuse) constructs that achieve a desired formatting effect on older browsers. They must be aware that these practices cause accessibility problems and must consider whether the formatting effect is so critical as to warrant making the document inaccessible to some users.

At the other extreme, content developers must not sacrifice appropriate markup because a certain browser or assistive technology does not process it correctly. For example, it is appropriate to use the TABLE element in HTML to mark up *tabular information* even though some older screen readers may not handle side-by-side text correctly (refer to checkpoint 10.3). Using TABLE correctly and creating tables that transform gracefully (refer to guideline 5) makes it possible for software to render tables other than as two-dimensional grids.

Checkpoints

3.1 When an appropriate markup language exists, use markup rather than images to convey information. [**Priority 2**]

For example, use MathML to mark up mathematical equations, and style sheets to format text and control layout. Also, avoid using images to represent text—use text and *style sheets* instead. Refer also to guideline 6 and guideline 11.

3.2 Create documents that validate to published formal grammars. [**Priority 2**]

For example, include a document type declaration at the beginning of a document that refers to a published DTD (e.g., the strict HTML 4.0 DTD).

3.3 Use style sheets to control layout and presentation. [**Priority 2**]

For example, use the CSS 'font' property instead of the HTML FONT element to control font styles.

3.4 Use relative rather than absolute units in markup language attribute values and style sheet property values. [**Priority 2**]

For example, in CSS, use 'em' or percentage lengths rather than 'pt' or 'cm', which are absolute units. If absolute units are used, validate that the rendered content is usable (refer to the section on validation).

3.5 Use header elements to convey document structure and use them according to specification. [**Priority 2**]

For example, in HTML, use H2 to indicate a subsection of H1. Do not use headers for font effects.

3.6 Mark up lists and list items properly. [**Priority 2**]

For example, in HTML, nest OL, UL, and DL lists properly.

3.7 Mark up quotations. Do not use quotation markup for formatting effects such as indentation. [**Priority 2**]

For example, in HTML, use the Q and BLOCKQUOTE elements to markup short and longer quotations, respectively.

Guideline 4. Clarify natural language usage

Use markup that facilitates pronunciation or interpretation of abbreviated or foreign text.

When content developers mark up natural language changes in a document, speech synthesizers and braille devices can automatically switch to the new language, making the document more accessible to multilingual users. Content developers should identify the predominant *natural language* of a document's content (through markup or HTTP headers). Content developers should also provide expansions of abbreviations and acronyms.

In addition to helping assistive technologies, natural language markup allows search engines to find key words and identify documents in a desired language. Natural language markup also improves readability of the Web for all people, including those with learning disabilities, cognitive disabilities, or people who are deaf.

When abbreviations and natural language changes are not identified, they may be indecipherable when machine-spoken or brailled.

Checkpoints

4.1 Clearly identify changes in the natural language of a document's text and any *text equivalents* (e.g., captions). [**Priority 1**]

For example, in HTML use the "lang" attribute. In XML, use "xml:lang".

4.2 Specify the expansion of each abbreviation or acronym in a document where it first occurs. [**Priority 3**]

For example, in HTML, use the "title" attribute of the ABBR and ACRONYM elements. Providing the expansion in the main body of the document also helps document usability.

4.3 Identify the primary natural language of a document. [**Priority 3**]

For example, in HTML set the "lang" attribute on the HTML element. In XML, use "xml:lang". Server operators should configure servers to take advantage of HTTP content negotiation mechanisms ([RFC2068], section 14.13) so that clients can automatically retrieve documents of the preferred language.

Guideline 5. Create tables that transform gracefully.

Ensure that tables have necessary markup to be transformed by accessible browsers and other user agents.

Tables should be used to mark up truly *tabular information* ("data tables"). Content developers should avoid using them to lay out pages ("layout tables"). Tables for any use also present special problems to users of *screen readers* (refer to checkpoint 10.3).

Some *user agents* allow users to navigate among table cells and access header and other table cell information. Unless marked-up properly, these tables will not provide user agents with the appropriate information. (Refer also to guideline 3.)

The following checkpoints will directly benefit people who access a table through auditory means (e.g., a screen reader or an automobile-based personal computer) or who view only a portion of the page at a time (e.g., users with blindness or low vision using speech output or a *braille display*, or other users of devices with small displays, etc.).

Checkpoints

5.1 For data tables, identify row and column headers. [**Priority 1**]

For example, in HTML, use TD to identify data cells and TH to identify headers.

5.2 For data tables that have two or more logical levels of row or column headers, use markup to associate data cells and header cells. [**Priority 1**]

For example, in HTML, use THEAD, TFOOT, and TBODY to group rows, COL and COLGROUP to group columns, and the "axis", "scope", and "headers" attributes, to describe more complex relationships among data.

5.3 Do not use tables for layout unless the table makes sense when linearized. Otherwise, if the table does not make sense, provide an alternative equivalent (which may be a *linearized version*). [**Priority 2**]

Note *Once user agents* support style sheet positioning, tables should not be used for layout. Refer also to checkpoint 3.3.

5.4 If a table is used for layout, do not use any structural markup for the purpose of visual formatting. [**Priority 2**]

For example, in HTML do not use the TH element to cause the content of a (non-table header) cell to be displayed centered and in bold.

5.5 Provide summaries for tables. [**Priority 3**]
For example, in HTML, use the "summary" attribute of the TABLE element.

5.6 Provide abbreviations for header labels. [**Priority 3**]
For example, in HTML, use the "abbr" attribute on the TH element. Refer also to checkpoint 10.3.

Guideline 6. Ensure that pages featuring new technologies transform gracefully.

Ensure that pages are accessible even when newer technologies are not supported or are turned off.

Although content developers are encouraged to use new technologies that solve problems raised by existing technologies, they should know how to make their pages still work with older browsers and people who choose to turn off features.

Checkpoints

6.1 Organize documents so they may be read without style sheets. For example, when an HTML document is rendered without associated style sheets, it must still be possible to read the document. [**Priority 1**]
When content is organized logically, it will be rendered in a meaningful order when style sheets are turned off or not supported.

6.2 Ensure that equivalents for dynamic content are updated when the dynamic content changes. [**Priority 1**]

6.3 Ensure that pages are usable when scripts, applets, or other programmatic objects are turned off or not supported. If this is not possible, provide equivalent information on an alternative accessible page. [**Priority 1**]
For example, ensure that links that trigger scripts work when scripts are turned off or not supported (e.g., do not use "javascript:" as the link target). If it is not possible to make the page usable without scripts, provide a text equivalent with the NOSCRIPT element, or use a server-side script instead of a client-side script, or provide an alternative accessible page as per checkpoint 11.4. Refer also to guideline 1.

6.4 For scripts and applets, ensure that event handlers are input device-independent. [**Priority 2**]
Refer to the definition of device independence.

6.5 Ensure that dynamic content is accessible or provide an alternative presentation or page. [**Priority 2**]

For example, in HTML, use NOFRAMES at the end of each frameset. For some applications, server-side scripts may be more accessible than client-side scripts. Refer also to checkpoint 11.4.

Guideline 7. Ensure user control of time-sensitive content changes.

Ensure that moving, blinking, scrolling, or auto-updating objects or pages may be paused or stopped.

Some people with cognitive or visual disabilities are unable to read moving text quickly enough or at all. Movement can also cause such a distraction that the rest of the page becomes unreadable for people with cognitive disabilities. *Screen readers* are unable to read moving text. People with physical disabilities might not be able to move quickly or accurately enough to interact with moving objects.

Note. All of the following checkpoints involve some content developer responsibility *until user agents* provide adequate feature control mechanisms.

Checkpoints

7.1 *Until user agents* allow users to control flickering, avoid causing the screen to flicker. [**Priority 1**]

Note People with photosensitive epilepsy can have seizures triggered by flickering or flashing in the 4 to 59 flashes per second (Hertz) range with a peak sensitivity at 20 flashes per second as well as quick changes from dark to light (like strobe lights).

7.2 *Until user agents* allow users to control blinking, avoid causing content to blink (i.e., change presentation at a regular rate, such as turning on and off). [**Priority 2**]

7.3 *Until user agents* allow users to freeze moving content, avoid movement in pages. [**Priority 2**]

When a page includes moving content, provide a mechanism within a script or applet to allow users to freeze motion or updates. Using style sheets with scripting to create movement allows users to turn off or override the effect more easily. Refer also to guideline 8.

7.4 *Until user agents* provide the ability to stop the refresh, do not create periodically auto-refreshing pages. [**Priority 2**]

For example, in HTML, don't cause pages to auto-refresh with "HTTP-EQUIV=refresh" until user agents allow users to turn off the feature.

7.5 *Until user agents* provide the ability to stop auto-redirect, do not use markup to redirect pages automatically. Instead, configure the server to perform redirects. [**Priority 2**]

Note The BLINK and MARQUEE elements are not defined in any W3C HTML specification and should not be used. Refer also to guideline 11.

Guideline 8. Ensure direct accessibility of embedded user interfaces.

Ensure that the user interface follows principles of accessible design: device-independent access to functionality, keyboard operability, self-voicing, etc.

When an embedded object has its "own interface", the interface—like the interface to the browser itself—must be accessible. If the interface of the embedded object cannot be made accessible, an alternative accessible solution must be provided.

Note For information about accessible interfaces, please consult the User Agent Accessibility Guidelines ([WAI-USERAGENT]) and the Authoring Tool Accessibility Guidelines ([WAI-AUTOOL]).

Checkpoint:

8.1 Make programmatic elements such as scripts and applets directly accessible or compatible with assistive technologies [**Priority 1 if functionality is *important* and not presented elsewhere, otherwise Priority 2.**]

Refer also to guideline 6.

Guideline 9. Design for device-independence.

Use features that enable activation of page elements via a variety of input devices.

Device-independent access means that the user may interact with the user agent or document with a preferred input (or output) device—mouse, keyboard, voice, head wand, or other. If, for example, a form control can only be activated with a mouse or other pointing device, someone who is using the page without sight, with voice input, or with a keyboard or who is using some other non-pointing input device will not be able to use the form.

Note. Providing text equivalents for image maps or images used as links makes it possible for users to interact with them without a pointing device. Refer also to guideline 1.

Generally, pages that allow keyboard interaction are also accessible through speech input or a command line interface.

Checkpoints

9.1 Provide client-side image maps instead of server-side image maps except where the regions cannot be defined with an available geometric shape. [**Priority 1**]

Refer also to checkpoint 1.1, checkpoint 1.2, and checkpoint 1.5.

9.2 Ensure that any element that has its own interface can be operated in a device-independent manner. [**Priority 2**]

Refer to the definition of device independence. Refer also to guideline 8.

9.3 For scripts, specify logical event handlers rather than device-dependent event handlers. [**Priority 2**]

9.4 Create a logical tab order through links, form controls, and objects. [**Priority 3**]

For example, in HTML, specify tab order via the "tabindex" attribute or ensure a logical page design.

9.5 Provide keyboard shortcuts to important links (including those in *client-side image maps*), form controls, and groups of form controls. [**Priority 3**]

For example, in HTML, specify shortcuts via the "accesskey" attribute.

Guideline 10. Use interim solutions.

Use interim accessibility solutions so that assistive technologies and older browsers will operate correctly.

For example, older browsers do not allow users to navigate to empty edit boxes. Older screen readers read lists of consecutive links as one link. These active elements are therefore difficult or impossible to access. Also, changing the current window or popping up new windows can be very disorienting to users who cannot see that this has happened.

Note The following checkpoints apply *until user agents* (including *assistive technologies*) address these issues. These checkpoints are classified as "interim", meaning that the Web Content Guidelines Working Group considers them to be valid and necessary to Web accessibility *as of the publication of this document*. However, the Working Group does not expect these checkpoints to be necessary in the future, once Web technologies have incorporated anticipated features or capabilities.

Checkpoints

10.1 *Until user agents* allow users to turn off spawned windows, do not cause pop-ups or other windows to appear and do not change the current window without informing the user. [Priority 2]
For example, in HTML, avoid using a frame whose target is a new window.

10.2 *Until user agents* support explicit associations between labels and form controls, for all form controls with implicitly associated labels, ensure that the label is properly positioned. [Priority 2]
The label must immediately precede its control on the same line (allowing more than one control/label per line) or be in the line preceding the control (with only one label and one control per line). Refer also to checkpoint 12.4.

10.3 *Until user agents* (including assistive technologies) render side-by-side text correctly, provide a linear text alternative (on the current page or some other) for *all* tables that lay out text in parallel, word-wrapped columns. [Priority 3]

Note Please consult the definition of *linearized table*. This checkpoint benefits people with *user agents* (such as some *screen readers*) that are unable to handle blocks of text presented side-by-side; the checkpoint should not discourage content developers from using tables to represent *tabular information*.

10.4 *Until user agents* handle empty controls correctly, include default, place-holding characters in edit boxes and text areas. [Priority 3]
For example, in HTML, do this for TEXTAREA and INPUT.

10.5 *Until user agents* (including assistive technologies) render adjacent links distinctly, include non-link, printable characters (surrounded by spaces) between adjacent links. [Priority 3]

Guideline 11. Use W3C technologies and guidelines.

Use W3C technologies (according to specification) and follow accessibility guidelines. Where it is not possible to use a W3C technology, or doing so results in material that does not transform gracefully, provide an alternative version of the content that is accessible.

The current guidelines recommend W3C technologies (e.g., HTML, CSS, etc.) for several reasons.

- W3C technologies include "built-in" accessibility features.
- W3C specifications undergo early review to ensure that accessibility issues are considered during the design phase.
- W3C specifications are developed in an open, industry consensus process.

Many non-W3C formats (e.g., PDF, Shockwave, etc.) require viewing with either plug-ins or stand-alone applications. Often, these formats cannot be viewed or navigated with standard *user agents* (including *assistive technologies*). Avoiding non-W3C and non-standard features (proprietary elements, attributes, properties, and extensions) will tend to make pages more accessible to more people using a wider variety of hardware and software. When inaccessible technologies (proprietary or not) must be used, equivalent accessible pages must be provided.

Even when W3C technologies are used, they must be used in accordance with accessibility guidelines. When using new technologies, ensure that they transform gracefully (Refer also to guideline 6.).

Note Converting documents (from PDF, PostScript, RTF, etc.) to W3C markup languages (HTML, XML) does not always create an accessible document. Therefore, validate each page for accessibility and usability after the conversion process (refer to the section on validation). If a page does not readily convert, either revise the page until its original representation converts appropriately or provide an HTML or plain text version.

Checkpoints

11.1 Use W3C technologies when they are available and appropriate for a task and use the latest versions when supported. [**Priority 2**]

Refer to the list of references for information about where to find the latest W3C specifications and [WAI-UA-SUPPORT] for information about user agent support for W3C technologies.

11.2 Avoid deprecated features of W3C technologies. [**Priority 2**]

For example, in HTML, don't use the *deprecated* FONT element; use style sheets instead (e.g., the 'font' property in CSS).

11.3 Provide information so that users may receive documents according to their preferences (e.g., language, content type, etc.) [**Priority 3**]

Note   Use content negotiation where possible.

11.4 If, after best efforts, you cannot create an *accessible* page, provide a link to an alternative page that uses W3C technologies, is accessible, has *equivalent* information (or functionality), and is updated as often as the inaccessible (original) page. [**Priority 1**]

Note   Content developers should only resort to alternative pages when other solutions fail because alternative pages are generally updated less often than "primary" pages. An out-of-date page may be as frustrating as one that is inaccessible since, in both cases, the information presented on the original page is unavailable. Automatically generating alternative pages may lead to more frequent updates, but content developers must still be careful to ensure that generated pages always make sense, and that users are able to navigate a site by following links on primary pages, alternative pages, or both. Before resorting to an alternative page, reconsider the design of the original page; making it accessible is likely to improve it for all users.

Guideline 12. Provide context and orientation information.

Provide context and orientation information to help users understand complex pages or elements.

Grouping elements and providing contextual information about the relationships between elements can be useful for all users. Complex relationships between parts of a page may be difficult for people with cognitive disabilities and people with visual disabilities to interpret.

Checkpoints

12.1 Title each frame to facilitate frame identification and navigation. [**Priority 1**]

For example, in HTML use the "title" attribute on FRAME elements.

12.2 Describe the purpose of frames and how frames relate to each other if it is not obvious by frame titles alone. [**Priority 2**]

For example, in HTML, use "longdesc," or a *description link*.

12.3 Divide large blocks of information into more manageable groups where natural and appropriate. [**Priority 2**]

For example, in HTML, use OPTGROUP to group OPTION elements inside a SELECT; group form controls with FIELDSET and LEGEND; use nested lists where appropriate; use headings to structure documents, etc. Refer also to guideline 3.

12.4 Associate labels explicitly with their controls. [**Priority 2**]

For example, in HTML use LABEL and its "for" attribute.

Guideline 13. Provide clear navigation mechanisms.

Provide clear and consistent navigation mechanisms—orientation information, navigation bars, a site map, etc.—to increase the likelihood that a person will find what they are looking for at a site.

Clear and consistent navigation mechanisms are important to people with cognitive disabilities or blindness, and benefit all users.

Checkpoints

13.1 Clearly identify the target of each link. [**Priority 2**]

Link text should be meaningful enough to make sense when read out of context—either on its own or as part of a sequence of links. Link text should also be terse. For example, in HTML, write "Information about version 4.3" instead of "click here". In addition to clear link text, content developers may further clarify the target of a link with an informative link title (e.g., in HTML, the "title" attribute).

13.2 Provide metadata to add semantic information to pages and sites. [**Priority 2**]

For example, use RDF ([RDF]) to indicate the document's author, the type of content, etc.

Note Some HTML *user agents* can build navigation tools from document relations described by the HTML LINK element and "rel" or "rev" attributes (e.g., rel="next", rel="previous", rel="index", etc.). Refer also to checkpoint 13.5.

13.3 Provide information about the general layout of a site (e.g., a site map or table of contents). [**Priority 2**]

In describing site layout, highlight and explain available accessibility features.

13.4 Use navigation mechanisms in a consistent manner. [**Priority 2**]

13.5 Provide navigation bars to highlight and give access to the navigation mechanism. [**Priority 3**]

13.6 Group related links, identify the group (for user agents), and, *until user agents* do so, provide a way to bypass the group. [**Priority 3**]

13.7 If search functions are provided, enable different types of searches for different skill levels and preferences. [**Priority 3**]

13.8 Place distinguishing information at the beginning of headings, paragraphs, lists, etc. [**Priority 3**]

Note This is commonly referred to as "front-loading" and is especially helpful for people accessing information with serial devices such as speech synthesizers.

13.9 Provide information about document collections (i.e., documents comprising multiple pages.). [**Priority 3**]
   For example, in HTML specify document collections with the LINK element and the "rel" and "rev" attributes. Another way to create a collection is by building an archive (e.g., with zip, tar and gzip, stuffit, etc.) of the multiple pages. Note. The performance improvement gained by offline processing can make browsing much less expensive for people with disabilities who may be browsing slowly.

13.10 Provide a means to skip over multi-line ASCII art. [**Priority 3**]
   Refer to checkpoint 1.1 and the example of ascii art in the glossary.

Guideline 14. Ensure that documents are clear and simple.

Ensure that documents are clear and simple so they may be more easily understood.

Consistent page layout, recognizable graphics, and easy to understand language benefit all users. In particular, they help people with cognitive disabilities or who have difficulty reading. (However, ensure that images have text equivalents for people who are blind, have low vision, or for any user who cannot or has chosen not to view graphics. Refer also to guideline 1.)

Using clear and simple language promotes effective communication. Access to written information can be difficult for people who have cognitive or learning disabilities. Using clear and simple language also benefits people whose first language differs from your own, including those people who communicate primarily in sign language.

Checkpoints

14.1 Use the clearest and simplest language appropriate for a site's content. [**Priority 1**]

14.2 Supplement text with graphic or auditory presentations where they will facilitate comprehension of the page. [**Priority 3**]
    Refer also to guideline 1.

14.3 Create a style of presentation that is consistent across pages. [**Priority 3**]

Appendix A.—Validation

Validate accessibility with automatic tools and human review. Automated methods are generally rapid and convenient but cannot identify all accessibility issues. Human review can help ensure clarity of language and ease of navigation.
    Begin using validation methods at the earliest stages of development. Accessibility issues identified early are easier to correct and avoid.
    Following are some important validation methods, discussed in more detail in the section on validation in the Techniques Document.

1. Use an automated accessibility tool and browser validation tool. Please note that software tools do not address all accessibility issues, such as the meaningfulness of link text, the applicability of a *text equivalent*, etc.
2. Validate syntax (e.g., HTML, XML, etc.).
3. Validate style sheets (e.g., CSS).
4. Use a text-only browser or emulator.
5. Use multiple graphic browsers, with:
   - sounds and graphics loaded,
   - graphics not loaded,
   - sounds not loaded,
   - no mouse,
   - frames, scripts, style sheets, and applets not loaded
6. Use several browsers, old and new.

7. Use a self-voicing browser, a screen reader, magnification software, a small display, etc.

8. Use spell and grammar checkers. A person reading a page with a speech synthesizer may not be able to decipher the synthesizer's best guess for a word with a spelling error. Eliminating grammar problems increases comprehension.

9. Review the document for clarity and simplicity. Readability statistics, such as those generated by some word processors may be useful indicators of clarity and simplicity. Better still, ask an experienced (human) editor to review written content for clarity. Editors can also improve the usability of documents by identifying potentially sensitive cultural issues that might arise due to language or icon usage.

10. Invite people with disabilities to review documents. Expert and novice users with disabilities will provide valuable feedback about accessibility or usability problems and their severity.

Appendix B.—Glossary

Accessible   Content is accessible when it may be used by someone with a disability.

Apple   A program inserted into a Web page.

Assistive technology   Software or hardware that has been specifically designed to assist people with disabilities in carrying out daily activities. Assistive technology includes wheelchairs, reading machines, devices for grasping, etc. In the area of Web Accessibility, common software-based assistive technologies include screen readers, screen magnifiers, speech synthesizers, and voice input software that operate in conjunction with graphical desktop browsers (among other *user agents*). Hardware assistive technologies include alternative keyboards and pointing devices.

ASCII art   ASCII art refers to text characters and symbols that are combined to create an image. For example ";-)" is the smiley emoticon. The following is an ascii figure showing the relationship between flash frequency and photoconvulsive response in patients with eyes open and closed [skip over ascii figure or consult a description of chart]:

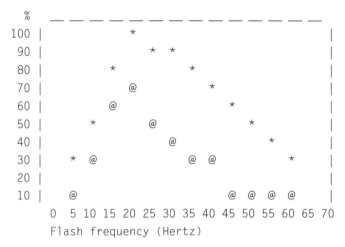

Authoring tool   HTML editors, document conversion tools, tools that generate Web content from databases are all authoring tools. Refer to the "Authoring Tool Accessibility Guidelines" ([WAI-AUTOOLS]) for information about developing accessible tools.

Backward compatible   Design that continues to work with earlier versions of a language, program, etc.

Braille   Braille uses six raised dots in different patterns to represent letters and numbers to be read by people who are blind with their fingertips.

A **braille display,** commonly referred to as a "dynamic braille display," raises or lowers dot patterns on command from an electronic device, usually a computer. The result is a line of braille that can change from moment to moment. Current dynamic braille displays range in size from one cell (six or eight dots) to an eighty-cell line, most having between twelve and twenty cells per line.

Content developer   Someone who authors Web pages or designs Web sites.

Deprecated   A deprecated element or attribute is one that has been outdated by newer constructs. Deprecated elements may become obsolete in future versions of HTML. The index of HTML elements and attributes in the Techniques Document indicates which elements and attributes are deprecated in HTML 4.0.

Authors should avoid using deprecated elements and attributes. User agents should continue to support for reasons of backward compatibility.

Device independent Users must be able to interact with a user agent (and the document it renders) using the supported input and output devices of their choice and according to their needs. Input devices may include pointing devices, keyboards, braille devices, head wands, microphones, and others. Output devices may include monitors, speech synthesizers, and braille devices.

Please note that "device-independent support" does not mean that user agents must support every input or output device. User agents should offer redundant input and output mechanisms for those devices that are supported. For example, if a user agent supports keyboard and mouse input, users should be able to interact with all features using either the keyboard or the mouse.

Document Content, Structure, and Presentation The content of a document refers to what it says to the user through natural language, images, sounds, movies, animations, etc. The structure of a document is how it is organized logically (e.g., by chapter, with an introduction and table of contents, etc.).

An *element* (e.g., P, STRONG, BLOCKQUOTE in HTML) that specifies document structure is called a structural element. The presentation of a document is how the document is rendered (e.g., as print, as a two-dimensional graphical presentation, as an text-only presentation, as synthesized speech, as braille, etc.) An *element* that specifies document presentation (e.g., B, FONT, CENTER) is called a presentation element.

Consider a document header, for example. The content of the header is what the header says (e.g., "Sailboats"). In HTML, the header is a structural element marked up with, for example, an H2 element. Finally, the presentation of the header might be a bold block text in the margin, a centered line of text, a title spoken with a certain voice style (like an aural font), etc.

Dynamic HTML (DHTML) DHTML is the marketing term applied to a mixture of standards including HTML, *style sheets*, the Document Object Model [DOM1] and scripting. However, there is no W3C specification that formally defines DHTML. Most guidelines may be applicable to applications using DHTML, however the following guidelines focus on issues related to scripting and style sheets: guideline 1, guideline 3, guideline 6, guideline 7, and guideline 9.

Element This document uses the term "element" both in the strict SGML sense (an element is a syntactic construct) and more generally to mean a type of content (such as video or sound) or a logical construct (such as a header or list). The second sense emphasizes that a guideline inspired by HTML could easily apply to another markup language.

Note that some (SGML) elements have content that is rendered (e.g., the P, LI, or TABLE elements in HTML), some are replaced by external content (e.g., IMG), and some affect processing (e.g., STYLE and SCRIPT cause information to be processed by a style sheet or script engine). An element that causes text characters to be part of the document is called a text element.

Equivalent Content is "equivalent" to other content when both fulfill essentially the same function or purpose upon presentation to the user. In the context of this document, the equivalent must fulfill essentially the same function for the person with a disability (at least insofar as is feasible, given the nature of the disability and the state of technology), as the primary content does for the person without any disability. For example, the text "The Full Moon" might convey the same information as an image of a full moon when presented to users. Note that equivalent information focuses on **fulfilling the same function**. If the image is part of a link and understanding the image is crucial to guessing the link target, an equivalent must also give users an idea of the link target. Providing equivalent information for inaccessible content is one of the primary ways authors can make their documents accessible to people with disabilities.

As part of fulfilling the same function of content an equivalent may involve a description of that content (i.e., what the content looks like or sounds like). For example, in order for users to understand the information conveyed by a complex chart, authors should describe the visual information in the chart.

Since text content can be presented to the user as synthesized speech, braille, and visually-displayed text, these guidelines require **text equivalents** for graphic and audio information. Text equivalents must be written so that they convey all essential content. **Non-text equivalents** (e.g., an auditory description of a visual presentation, a video of a person telling a story using sign language as an equivalent for a written story, etc.) also improve accessibility for people who cannot access visual information or written text, including many individuals with blindness, cognitive disabilities, learning disabilities, and deafness.

Equivalent information may be provided in a number of ways, including through attributes (e.g., a text value for the "alt" attribute in HTML and SMIL), as part of element content (e.g., the OBJECT in HTML), as part of the document's prose, or via a linked document (e.g., designated by the "longdesc" attribute in HTML or a description link). Depending on the complexity of the equivalent, it may be necessary to combine techniques (e.g., use "alt" for an abbreviated equivalent, useful to familiar readers, in addition to "longdesc" for a link to more complete information, useful to first-time readers). The details of how and when to provide equivalent information are part of the Techniques Document ([TECHNIQUES]).

A **text transcript** is a text equivalent of audio information that includes spoken words and non-spoken sounds such as sound effects. A **caption** is a text transcript for the audio track of a video presentation that is synchronized with the video and audio tracks. Captions are generally rendered visually by being superimposed over the video, which benefits people who are deaf and hard-of-hearing, and anyone who cannot hear the audio (e.g., when in a crowded room). A **collated text transcript** combines (collates) captions with text descriptions of video information (descriptions of the actions, body language, graphics, and scene changes of the video track). These text equivalents make presentations accessible to people who are deaf-blind and to people who cannot play movies, animations, etc. It also makes the information available to search engines.

One example of a non-text equivalent is an **auditory description** of the key visual elements of a presentation. The description is either a prerecorded human voice or a synthesized voice (recorded or generated on the fly). The auditory description is synchronized with the audio track of the presentation, usually during natural pauses in the audio track. Auditory descriptions include information about actions, body language, graphics, and scene changes.

Image A graphical presentation.

Image map An image that has been divided into regions with associated actions. Clicking on an active region causes an action to occur.

When a user clicks on an active region of a client-side image map, the user agent calculates in which region the click occurred and follows the link associated with that region. Clicking on an active region of a server-side image map causes the coordinates of the click to be sent to a server, which then performs some action.

Content developers can make client-side image maps accessible by providing device-independent access to the same links associated with the image map's regions. Client-side image maps allow the user agent to provide immediate feedback as to whether or not the user's pointer is over an active region.

Important   Information in a document is important if understanding that information is crucial to understanding the document.

Linearized table   A table rendering process where the contents of the cells become a series of paragraphs (e.g., down the page) one after another. The paragraphs will occur in the same order as the cells are defined in the document source. Cells should make sense when read in order and should include structural elements (that create paragraphs, headers, lists, etc.) so the page makes sense after linearization.

Link text   The rendered text content of a link.

Natural Language   Spoken, written, or signed human languages such as French, Japanese, American Sign Language, and braille. The natural language of content may be indicated with the "lang" attribute in HTML ([HTML40], section 8.1) and the "xml:lang" attribute in XML ([XML], section 2.12).

Navigation Mechanism   A navigation mechanism is any means by which a user can navigate a page or site. Some typical mechanisms include:

navigation bars   A navigation bar is a collection of links to the most important parts of a document or site.

site maps   A site map provides a global view of the organization of a page or site.

tables of contents   A table of contents generally lists (and links to) the most important sections of a document.

Personal Digital Assistant (PDA)   A PDA is a small, portable computing device. Most PDAs are used to track personal data such as calendars, contacts, and electronic mail. A PDA is generally a handheld device with a small screen that allows input from various sources.

Screen magnifier   A software program that magnifies a portion of the screen, so that it can be more easily viewed. Screen magnifiers are used primarily by individuals with low vision.

Screen reader   A software program that reads the contents of the screen aloud to a user. Screen readers are used primarily by individuals who are blind. Screen readers can usually only read text that is printed, not painted, to the screen.

Style sheets A style sheet is a set of statements that specify presentation of a document. Style sheets may have three different origins: they may be written by content providers, created by users, or built into user agents. In CSS ([CSS2]), the interaction of content provider, user, and user agent style sheets is called the cascade.

Presentation markup is markup that achieves a stylistic (rather than structuring) effect such as the B or I elements in HTML. Note that the STRONG and EM elements are not considered presentation markup since they convey information that is independent of a particular font style.

Tabular information When tables are used to represent logical relationships among data—text, numbers, images, etc., that information is called "tabular information" and the tables are called "data tables". The relationships expressed by a table may be rendered visually (usually on a two-dimensional grid), aurally (often preceding cells with header information), or in other formats.

Until user agents ... In most of the checkpoints, content developers are asked to ensure the accessibility of their pages and sites. However, there are accessibility needs that would be more appropriately met by *user agents* (including *assistive technologies*). As of the publication of this document, not all user agents or assistive technologies provide the accessibility control users require (e.g., some user agents may not allow users to turn off blinking content, or some screen readers may not handle tables well). Checkpoints that contain the phrase "until user agents ..." require content developers to provide additional support for accessibility until most user agents readily available to their audience include the necessary accessibility features.

Note The W3C WAI Web site (refer to [WAI-UA-SUPPORT]) provides information about user agent support for accessibility features. Content developers are encouraged to consult this page regularly for updated information.

User agent Software to access Web content, including desktop graphical browsers, text browsers, voice browsers, mobile phones, multimedia players, plug-ins, and some software assistive technologies used in conjunction with browsers such as screen readers, screen magnifiers, and voice recognition software.

Acknowledgments

Web Content Guidelines Working Group Co-Chairs:
  Chuck Letourneau, Starling Access Services
  Gregg Vanderheiden, Trace Research and Development
W3C Team contacts:
  Judy Brewer and Daniel Dardailler
We wish to thank the following people who have contributed their time and valuable comments to shaping these guidelines:
  Harvey Bingham, Kevin Carey, Chetz Colwell, Neal Ewers, Geoff Freed, Al Gilman, Larry Goldberg, Jon Gunderson, Eric Hansen, Phill Jenkins, Leonard Kasday, George Kerscher, Marja-Riitta Koivunen, Josh Krieger, Scott Luebking, William Loughborough, Murray Maloney, Charles McCathieNevile, MegaZone (Livingston Enterprises), Masafumi Nakane, Mark Novak, Charles Oppermann, Mike Paciello, David Pawson, Michael Pieper, Greg Rosmaita, Liam Quinn, Dave Raggett, T.V. Raman, Robert Savellis, Jutta Treviranus, Steve Tyler, Jaap van Lelieveld, and Jason White
The original draft of this document is based on "The Unified Web Site Accessibility Guidelines" ([UWSAG]) compiled by the Trace R & D Center at the University of Wisconsin. That document includes a list of additional contributors.

References

For the latest version of any W3C specification please consult the list of W3C Technical Reports.

[CSS1] "CSS, level 1 Recommendation", B. Bos, H. Wium Lie, eds., 17 December 1996, revised 11 January 1999. The CSS1 Recommendation is: http://www.w3.org/TR/1999/REC-CSS1-19990111.
The latest version of CSS1 is available at: http://www.w3.org/TR/REC-CSS1.

[CSS2] "CSS, level 2 Recommendation", B. Bos, H. Wium Lie, C. Lilley, and I. Jacobs, eds., 12 May 1998. The CSS2 Recommendation is: http://www.w3.org/TR/1998/REC-CSS2-19980512.
The latest version of CSS2 is available at: http://www.w3.org/TR/REC-CSS2.

[DOM1] "Document Object Model (DOM) Level 1 Specification", V. Apparao, S. Byrne, M. Champion, S. Isaacs, I. Jacobs, A. Le Hors, G. Nicol, J. Robie, R. Sutor, C. Wilson, and L. Wood, eds., 1 October 1998. The DOM Level 1 Recommendation is: http://www.w3.org/TR/1998/

REC-DOM-Level-1-19981001.
The latest version of DOM Level 1 is available at: `http://www.w3.org/TR/REC-DOM-Level-1`.

[HTML40] "HTML 4.0 Recommendation", D. Raggett, A. Le Hors, and I. Jacobs, eds., 17 December 1997, revised 24 April 1998. The HTML 4.0 Recommendation is: `http://www.w3.org/TR/1998/REC-html40-19980424`. The latest version of HTML 4.0 is available at: `http://www.w3.org/TR/REC-html40`.

[HTML32] "HTML 3.2 Recommendation", D. Raggett, ed., 14 January 1997. The latest version of HTML 3.2 is available at: `http://www.w3.org/TR/REC-html32`.

[MATHML] "Mathematical Markup Language", P. Ion and R. Miner, eds., 7 April 1998. The MathML 1.0 Recommendation is: `http://www.w3.org/TR/1998/REC-MathML-19980407`.

[PNG] "PNG (Portable Network Graphics) Specification", T. Boutell, ed., T. Lane, contributing ed., 1 October 1996. The latest version of PNG 1.0 is: `http://www.w3.org/TR/REC-png`.

[RDF] "Resource Description Framework (RDF) Model and Syntax Specification", O. Lassila, R. Swick, eds., 22 February 1999. The RDF Recommendation is: `http://www.w3.org/TR/1999/REC-rdf-syntax-19990222`. The latest version of RDF 1.0 is available at: `http://www.w3.org/TR/REC-rdf-syntax`.

[RFC2068] "HTTP Version 1.1", R. Fielding, J. Gettys, J. Mogul, H. Frystyk Nielsen, and T. Berners-Lee, January 1997.

[SMIL] "Synchronized Multimedia Integration Language (SMIL) 1.0 Specification", P. Hoschka, ed., 15 June 1998. The SMIL 1.0 Recommendation is: `http://www.w3.org/TR/1998/REC-smil-19980615`. The latest version of SMIL 1.0 is available at: `http://www.w3.org/TR/REC-smil`.

[TECHNIQUES] "Techniques for Web Content Accessibility Guidelines 1.0", W. Chisholm, G. Vanderheiden, I. Jacobs, eds. This document explains how to implement the checkpoints defined in "Web Content Accessibility Guidelines 1.0". The latest draft of the techniques is available at: `http://www.w3.org/TR/WAI-WEBCONTENT-TECHS/`.

[WAI-AUTOOLS] "Authoring Tool Accessibility Guidelines", J. Treviranus, J. Richards, I. Jacobs, C. McCathieNevile, eds. The latest Working

Draft of these guidelines for designing accessible authoring tools is available at: http://www.w3.org/TR/WAI-AUTOOLS/.

[WAI-UA-SUPPORT] This page documents known support by user agents (including assistive technologies) of some accessibility features listed in this document. The page is available at: http://www.w3.org/WAI/Resources/WAI-UA-Support.

[WAI-USERAGENT] "User Agent Accessibility Guidelines", J. Gunderson and I. Jacobs, eds. The latest Working Draft of these guidelines for designing accessible user agents is available at: http://www.w3.org/TR/WAI-USER-AGENT/.

[WCAG-ICONS] Information about conformance icons for this document and how to use them is available at http://www.w3.org/WAI/WCAG1-Conformance.html.

[UWSAG] "The Unified Web Site Accessibility Guidelines", G. Vanderheiden, W. Chisholm, eds. The Unified Web Site Guidelines were compiled by the Trace R & D Center at the University of Wisconsin under funding from the National Institute on Disability and Rehabilitation Research (NIDRR), U.S. Dept. of Education. This document is available at: http://www.tracecenter.org/docs/html_guidelines/version8.htm.

[XML] "Extensible Markup Language (XML) 1.0.", T. Bray, J. Paoli, C.M. Sperberg-McQueen, eds., 10 February 1998. The XML 1.0 Recommendation is: http://www.w3.org/TR/1998/REC-xml-19980210. The latest version of XML 1.0 is available at: http://www.w3.org/TR/REC-xml.

B

Cascading Style Sheets Specification

Cascading Style Sheets

In spite of what people might tell you, HTML is a markup language created to define Web page structure — not presentation. However, shortly after the introduction of graphical Web browsers, determined individuals realized they could force presentation features using certain HTML coding elements and attributes. Compounding the problem further, Web browsers would process the HTML because the browser is concerned about the validity of the element's existence, not the validity of the element's presentation definition.

Soon site designers began implementing tables to position text and graphics on a page. Then HTML headers were coded to achieve a particular font style with no regard for HTML's defined hierarchy (H1, H2, H3, and so on). In other cases, elements were used to convey structure that could only be understood in a visual context (for example, tabular data was presented using the PRE element).

Each of these examples complicates the rendering and comprehension of a Web page for a person with a visual disability. Tables, for example, by their very nature (columnar) are difficult for a blind person who uses a screen reader to comprehend because screen readers read lines of information left-to-right, line by line. You can imagine how difficult it is for a person who relies on a voice-based user agent or screen reader to interpret that a Web site's table isn't a table of tabular information, but rather a means for laying out the presentation of that page. Navigation is often a nightmare.

True, recent advances in screen reader technology have helped. But this is only after the screen reader completely reformats the table (and subsequently the Web page) in a manner that can be understood by the blind Web surfer.

Cascading style sheets, or CSS, eliminate the confusion by enabling a Web site designer to define explicitly the style of HTML elements separate from the structured markup. By separating presentation from structure, specialized technologies used by people with disabilities can easily interpret the structure and render it properly to the user. At the same time, CSS is a more efficient way of defining a Web site's presentation layout and style. For example, you can create one style sheet that applies to all your Web pages. If you make a change to the style sheet, that change is reflected in all the pages.

CSS also provides two other major benefits: support of media types and aural presentation of Web pages. Now Web site designers can specify media types for specific device types that support individuals with disabilities, including Braille-tactile feedback devices and speech synthesizers.

Two W3C style sheet specifications exist today: CSS1 and CSS2. Browsers and other user agents support various levels of both, but no user agent supports complete implementations of either specification. You may have noticed that some browsers, including Opera and Internet Explorer, enable you to define your own style sheets. This enables you to override the HTML presentation of Web page and present page elements according to your personal preferences. For example, people who are color blind can define specific colors that are easier for them to view.

The sections that follow are extractions from the W3C's CSS2 recommendation. These sections cover the following three areas:

1. Recognized Media Types
2. Audio Rendering of Tables
3. Aural Style Sheets

All three sections provide relevant information to help you design and create more accessible Web sites for people with disabilities.

CSS2 Recognized Media Types

Using media types can help you design Web pages for assistive and accessible devices that support people with disabilities, including Braille and synthetic speech synthesizers. This section duplicates the CSS2 draft found at `http://www.w3.org/TR/REC-CSS2/media.html`.

7.3 Recognized media types

A CSS media type names a set of CSS properties. A user agent that claims to support a media type by name must implement all of the properties that apply to that media type.

The names chosen for CSS media types reflect target devices for which the relevant properties make sense. In the following list of CSS media types, the parenthetical descriptions are not normative. They only give a sense of what device the media type is meant to refer to.

All   Suitable for all devices.

aural   Intended for speech synthesizers. See the section on aural style sheets for details.

braille   Intended for braille tactile feedback devices.

embossed   Intended for paged braille printers.

handheld   Intended for handheld devices (typically small screen, monochrome, limited bandwidth).

print   Intended for paged, opaque material and for documents viewed on screen in print preview mode. Please consult the section on paged media for information about formatting issues that are specific to paged media.

projection   Intended for projected presentations, for example projectors or print to transparencies. Please consult the section on paged media for information about formatting issues that are specific to paged media.

screen   Intended primarily for color computer screens.

tty Intended for media using a fixed-pitch character grid, such as tele-types, terminals, or portable devices with limited display capabilities. Authors should not use pixel units with the "tty" media type.

tv Intended for television-type devices (low resolution, color, lim-ited-scrollability screens, sound available).

Media type names are case-insensitive.

Due to rapidly changing technologies, CSS2 does not specify a definitive list of media types that may be values for @media.

Note Future versions of CSS may extend this list. Authors should not rely on media type names that are not yet defined by a CSS specification.

7.3.1 Media groups

Each CSS property definition specifies the media types for which theprop-erty must be implemented by a conforming user agent. Since properties gen-erally apply to several media, the "Applies to media" section of each property definition lists media groups rather than individual media types. Each property applies to all media types in the media groups listed in its def-inition.

CSS2 defines the following media groups.

Continuous or Paged "Both" means that the property in question applies to both media groups.

Visual, Aural, or Tactile

Grid (for character grid devices) or Bitmap "Both" means that the property in question applies to both media groups.

Interactive (for devices that allow user interaction) or Static (for those that don't) "Both" means that the property in question applies to both media groups.

all (includes all media types)

The following table shows the relationships between media groups and media types

Relationship between media groups and media types

Media Types	Media Groups			
	continuous/ paged	visual/aural/ tactile	grid/ bitmap	interactive/ static
aural	continuous	aural	N/A	both
braille	continuous	tactile	grid	both
emboss	paged	tactile	grid	both
handheld	both	visual	both	both
print	paged	visual	bitmap	static
projection	paged	visual	bitmap	static
screen	continuous	visual	bitmap	both
tty	continuous	visual	grid	both
tv	both	visual, aural	bitmap	both

CSS2 Audio Rendering of Tables

CSS2 includes a provision for marking up the presentation of tables for an audio-based interface (where date is rendered using voice output). This section duplicates the CSS2 draft found at `http://www.w3.org/TR/REC-CSS2/tables.html#q21`.

17.7 Audio rendering of tables

When a table is spoken by a speech generator, the relation between the data cells and the header cells must be expressed in a different way than by horizontal and vertical alignment. Some speech browsers may allow a user to move around in the 2-dimensional space, thus giving them the opportunity to map out the spatially represented relations. When that is not possible, the style sheet must specify at which points the headers are spoken.

17.7.1 Speaking headers: the 'speak-header' property

Value:	once I always I inherit
Initial:	once
Applies to:	elements that have table header information
Inherited:	yes
Percentages:	N/A
Media:	aural

This property specifies whether table headers are spoken before every cell, or only before a cell when that cell is associated with a different header than the previous cell. Values have the following meanings.

once   The header is spoken one time, before a series of cells.

always   The header is spoken before every pertinent cell.

Each document language may have different mechanisms that allow authors to specify headers. For example, in HTML 4.0 ([HTML40]), it is possible to specify header information with three different attributes (headers, scope, and axis), and the specification gives an algorithm for determining header information when these attributes have not been specified.

Figure B.1   **Image of a table with header cells that are not in the same column or row as the data they apply to.**

Travel Expense Report

	Meals	Hotels	Transport	subtotals
San Jose				
25-Aug-97	37.74	112.00	45.00	
26-Aug-97	27.28	112.00	45.00	
subtotals	65.02	224.00	90.00	379.02
Seattle				
27-Aug-97	96.25	109.00	36.00	
28-Aug-97	35.00	109.00	36.00	
subtotals	131.25	218.00	72.00	421.25
Totals	196.27	442.00	162.00	**800.27**

This HTML example presents the money spent on meals, hotels and transport in two locations (San Jose and Seattle) for successive days. Conceptually, you can think of the table in terms of a n-dimensional space. The headers of this space are: location, day, category and subtotal. Some cells define marks along an axis while others give money spent at points within this space. The markup for this table is as follows.

```
<TABLE>
<CAPTION>Travel Expense Report</CAPTION>
<TR>
      <TH></TH>
      <TH>Meals</TH>
      <TH>Hotels</TH>
      <TH>Transport</TH>
      <TH>subtotal</TH>
</TR>
<TR>
      <TH id="san-jose" axis="san-jose">San Jose</TH>
</TR>
<TR>
      <TH headers="san-jose">25-Aug-97</TH>
      <TD>37.74</TD>
      <TD>112.00</TD>
      <TD>45.00</TD>
      <TD></TD>
</TR>
<TR>
      <TH headers="san-jose">26-Aug-97</TH>
      <TD>27.28</TD>
       <TD>112.00</TD>
        <TD>45.00</TD>
          <TD></TD>
</TR>
<TR>
      <TH headers="san-jose">subtotal</TH>
      <TD>65.02</TD>
       <TD>224.00</TD>
        <TD>90.00</TD>
```

```
            <TD>379.02</TD>
    </TR>
    <TR>
        <TH id="seattle" axis="seattle">Seattle</TH>
        </TR>
    <TR>
        <TH headers="seattle">27-Aug-97</TH>
        <TD>96.25</TD>
        <TD>109.00</TD>
        <TD>36.00</TD>
        <TD></TD>
    </TR>
    <TR>
        <TH headers="seattle">28-Aug-97</TH>
        <TD>35.00</TD>
        <TD>109.00</TD>
        <TD>36.00</TD>
        <TD></TD>
    </TR>
    <TR>
        <TH headers="seattle">subtotal</TH>
        <TD>131.25</TD>
        <TD>218.00</TD>
        <TD>72.00</TD>
        <TD>421.25</TD>
    </TR>
    <TR>
        <TH>Totals</TH>
        <TD>196.27</TD>
        <TD>442.00</TD>
        <TD>162.00</TD>
        <TD>800.27</TD>
    </TR>
    </TABLE>
```

By providing the data model in this way, authors make it possible for speech enabled browsers to explore the table in rich ways, e.g., each cell could be spoken as a list, repeating the applicable headers before each data cell.

```
San Jose, 25-Aug-97, Meals:  37.74
San Jose, 25-Aug-97, Hotels:  112.00
San Jose, 25-Aug-97, Transport:  45.00
```

The browser could also speak the headers only when they change.

```
San Jose, 25-Aug-97, Meals: 37.74
         Hotels: 112.00
         Transport: 45.00
    26-Aug-97, Meals: 27.28
         Hotels: 112.00
```

CSS2 Aural Cascading Style Sheets

Aural Cascading Style Sheets (ACSS) define the style properties for aural presentation of a Web document. For example, ACSS can be constructed to control voice characteristics of speech synthesizers including volume, pitch, and speed. This is one way of enhancing Web content for a blind or print impaired Web user, whose primary mode for accessing the Web requires voice output.

ACSS uses a combination of speech synthesis and auditory icons (earcons). A browser supporting ACSS needn't support the visual presentation, although that redundant presentation is useful for low-vision users, and for those whose equipment has no visual component.

ACSS is a subset of the current CSS2 working draft recommendation produced by the World Wide Web Consortium (W3C). This section duplicates the ACSS working draft found at `http://www.w3.org/TR/REC-CSS2/aural.html`.

19. Aural style sheets

19.1 Introduction to aural style sheets

The aural rendering of a document, already commonly used by the blind and print-impaired communities, combines speech synthesis and "auditory icons." Often such aural presentation occurs by converting the document to plain text and feeding this to a screen reader — software or hardware that simply reads all the characters on the screen. This results in less effective presentation than would be the case if the document structure were retained. Style sheet properties for aural presentation may be used together with visual properties (mixed media) or as an aural alternative to visual presentation.

Besides the obvious accessibility advantages, there are other large markets for listening to information, including in-car use, industrial and medical documentation systems (intranets), home entertainment, and to help users learning to read or who have difficulty reading.

When using aural properties, the canvas consists of a three-dimensional physical space (sound surrounds) and a temporal space (one may specify sounds before, during, and after other sounds). The CSS properties also allow authors to vary the quality of synthesized speech (voice type, frequency, inflection, etc.).

Example(s):

```
H1, H2, H3, H4, H5, H6 {
        voice-family: paul;
        stress: 20;
        richness: 90;
        cue-before: url("ping.au")
}
P.heidi { azimuth: center-left
P.peter { azimuth: right }
P.goat  { volume: x-soft }
```

This will direct the speech synthesizer to speak headers in a voice (a kind of "audio font") called paul, on a flat tone, but in a very rich voice. Before speaking the headers, a sound sample will be played from the given URL. Paragraphs with class heidi will appear to come from front left (if the sound system is capable of spatial audio), and paragraphs of class peter from the right. Paragraphs with class goat will be very soft.

19.2 Volume properties: 'volume'

Value:	<number>	<percentage>	silent	x-soft	soft	medium	loud	x-loud	inherit
Initial:	medium
Applies to:	all elements
Inherited:	yes
Percentages:	refer to inherited value
Media:	aural

Volume refers to the median volume of the waveform. In other words, a highly inflected voice at a volume of 50 might peak well above that. The overall values are likely to be human adjustable for comfort, for example with a physical volume control (which would increase both the 0 and 100 values proportionately); what this property does is adjust the dynamic range.

Values have the following meanings.

<number> Any number between '0' and '100'. '0' represents the minimum audible volume level and 100 corresponds to the maximum comfortable level.

<percentage> Percentage values are calculated relative to the inherited value, and are then clipped to the range '0' to '100'.

silent No sound at all. The value '0' does not mean the same as 'silent'.

x-soft Same as '0'.

soft Same as '25'.

medium Same as '50'.

loud Same as '75'.

x-loud Same as '100'.

User agents should allow the values corresponding to '0' and '100' to be set by the listener. No one setting is universally applicable; suitable values depend on the equipment in use (speakers, headphones), the environment (in car, home theater, library) and personal preferences. Some examples:

- A browser for in-car use has a setting for when there is lots of background noise. '0' would map to a fairly high level and '100' to a quite high level. The speech is easily audible over the road noise but the overall dynamic range is compressed. Cars with better insulation might allow a wider dynamic range.

- Another speech browser is being used in an apartment, late at night, or in a shared study room. '0' is set to a very quiet level and '100' to a fairly quiet level, too. As with the first example, there is a low slope; the dynamic range is reduced. The actual volumes are low here, whereas they were high in the first example.

- In a quiet and isolated house, an expensive hi-fi home theater setup. '0' is set fairly low and '100' to quite high; there is wide dynamic range.

The same author style sheet could be used in all cases, simply by mapping the '0' and '100' points suitably at the client side.

19.3 Speaking properties: 'speak'

Value:	normal	none	spell-out	inherit
Initial:	normal
Applies to:	all elements
Inherited:	yes
Percentages:	N/A
Media:	aural

This property specifies whether text will be rendered aurally and if so, in what manner (somewhat analogous to the 'display' property). The possible values are:

none Suppresses aural rendering so that the element requires no time to render. Note, however, that descendants may override this value and will be spoken. (To be sure to suppress rendering of an element and its descendants, use the 'display' property).

normal Uses language-dependent pronunciation rules for rendering an element and its children.

spell-out   Spells the text one letter at a time (useful for acronyms and abbreviations).

Note the difference between an element whose 'volume' property has a value of 'silent' and an element whose 'speak' property has the value 'none'. The former takes up the same time as if it had been spoken, including any pause before and after the element, but no sound is generated. The latter requires no time and is not rendered (though its descendants may be).

19.4 Pause properties

'pause-before'

Value:	<time> | <percentage> | inherit
Initial:	depends on user agent
Applies to:	all elements
Inherited:	no
Percentages:	see prose
Media:	aural

'pause-after'

Value:	<time> | <percentage> | inherit
Initial:	depends on user agent
Applies to:	all elements
Inherited:	no
Percentages:	see prose
Media:	aural

These properties specify a pause to be observed before (or after) speaking an element's content. Values have the following meanings.

<time>   Expresses the pause in absolute time units (seconds and milliseconds).

<percentage>   Refers to the inverse of the value of the 'speech-rate' property. For example, if the speech-rate is 120 words per minute (i.e., a word takes half a second, or 500ms) then a 'pause-before' of 100% means a pause of 500 ms and a 'pause-before' of 20% means 100ms.

The pause is inserted between the element's content and any 'cue-before' or 'cue-after' content.

Authors should use relative units to create more robust style sheets in the face of large changes in speech-rate.

'pause'

Value:	[[<time> | <percentage>]{1,2}] | inherit
Initial:	depends on user agent
Applies to:	all elements
Inherited:	no
Percentages:	see descriptions of 'pause-before' and 'pause-after'
Media:	aural

The 'pause' property is a shorthand for setting 'pause-before' and 'pause-after'. If two values are given, the first value is 'pause-before' and the second is 'pause-after'. If only one value is given, it applies to both properties.

Example(s):

```
H1 { pause: 20ms } /* pause-before: 20ms; pause-after: 20ms */
H2 { pause: 30ms 40ms } /* pause-before: 30ms; pause-after: 40ms */
H3 { pause-after: 10ms } /* pause-before: ?; pause-after: 10ms */
```

19.5 Cue properties

'cue-before'

Value:	<uri> | none | inherit
Initial:	none
Applies to:	all elements
Inherited:	no
Percentages:	N/A
Media:	aural

'cue-after'

Value:	<uri>	none	inherit
Initial:	none
Applies to:	all elements
Inherited:	no
Percentages:	N/A
Media:	aural

Auditory icons are another way to distinguish semantic elements. Sounds may be played before and/or after the element to delimit it. Values have the following meanings.

<uri> The URI must designate an auditory icon resource. If the URI resolves to something other than an audio file, such as an image, the resource should be ignored and the property treated as if it had the value 'none'.

none No auditory icon is specified.

Example(s):

```
A {cue-before: url("bell.aiff"); cue-after: url("dong.wav") }
H1 {cue-before: url("pop.au"); cue-after: url("pop.au") }
```

'cue'

Value:	[<'cue-before'> || <'cue-after'>]	inherit
Initial:	not defined for shorthand properties
Applies to:	all elements
Inherited:	no
Percentages:	N/A
Media:	aural

The 'cue' property is a shorthand for setting 'cue-before' and 'cue-after'. If two values are given, the first value is 'cue-before' and the second is 'cue-after'. If only one value is given, it applies to both properties.

The following two rules are equivalent:

```
H1 {cue-before: url("pop.au"); cue-after: url("pop.au") }
H1 {cue: url("pop.au") }
```

If a user agent cannot render an auditory icon (e.g., the user's environment does not permit it), we recommend that it produce an alternative cue (e.g., popping up a warning, emitting a warning sound, etc.)

Please see the sections on the :before and :after pseudo-elements for information on other content generation techniques.

19.6 Mixing properties: 'play-during'

Value:	<uri> mix? repeat?	auto	none	inherit
Initial:	auto
Applies to:	all elements
Inherited:	no
Percentages:	N/A
Media:	aural

Similar to the 'cue-before' and 'cue-after' properties, this property specifies a sound to be played as a background while an element's content is spoken. Values have the following meanings.

<uri>   The sound designated by this <uri> is played as a background while the element's content is spoken.

mix   When present, this keyword means that the sound inherited from the parent element's 'play-during' property continues to play and the sound designated by the <uri> is mixed with it. If 'mix' is not specified the element's background sound replaces the parent's.

repeat   When present, this keyword means that the sound will repeat if it is too short to fill the entire duration of the element. Otherwise, the sound plays once and then stops. This is similar to the 'background-repeat' property. If the sound is too long for the element, it is clipped once the element has been spoken.

auto The sound of the parent element continues to play (it is not restarted, which would have been the case if this property had been inherited).

none This keyword means that there is silence. The sound of the parent element (if any) is silent during the current element and continues after the current element.

Example(s):

```
BLOCKQUOTE.sad { play-during: url("violins.aiff") }
BLOCKQUOTE Q   { play-during: url("harp.wav") mix }
SPAN.quiet     { play-during: none }
```

19.7 Spatial properties

Spatial audio is an important stylistic property for aural presentation. It provides a natural way to tell several voices apart, as in real life (people rarely all stand in the same spot in a room). Stereo speakers produce a lateral sound stage. Binaural headphones or the increasingly popular 5-speaker home theater setups can generate full surround sound, and multi-speaker setups can create a true three-dimensional sound stage. VRML 2.0 also includes spatial audio, which implies that in time consumer-priced spatial audio hardware will become more widely available.

'azimuth'

Value:	<angle> | [[left-side | far-left | left | center-left | center | center-right | right | far-right | right-side] || behind] | leftwards | rightwards | inherit
Initial:	center
Applies to:	all elements
Inherited:	yes
Percentages:	N/A
Media:	aural

Values have the following meanings.

<angle> Position is described in terms of an angle within the range '-360deg' to '360deg'. The value '0deg' means directly ahead in the center of the sound stage. '90deg' is to the right, '180deg' behind, and '270deg' (or, equivalently and more conveniently, '-90deg') to the left.

left-side   Same as '270deg'. With 'behind', '270deg'.

far-left   Same as '300deg'. With 'behind', '240deg'.

left   Same as '320deg'. With 'behind', '220deg'.

center-left   Same as '340deg'. With 'behind', '200deg'.

center   Same as '0deg'. With 'behind', '180deg'.

center-right   Same as '20deg'. With 'behind', '160deg'.

right   Same as '40deg'. With 'behind', '140deg'.

far-right   Same as '60deg'. With 'behind', '120deg'.

right-side   Same as '90deg'. With 'behind', '90deg'.

leftwards   Moves the sound to the left, relative to the current angle. More precisely, subtracts 20 degrees. Arithmetic is carried out modulo 360 degrees. Note that `leftwards` is more accurately described as "turned counter-clockwise," since it always subtracts 20 degrees, even if the inherited azimuth is already behind the listener (in which case the sound actually appears to move to the right).

rightwards   Moves the sound to the right, relative to the current angle. More precisely, adds 20 degrees. See 'leftwards' for arithmetic.

This property is most likely to be implemented by mixing the same signal into different channels at differing volumes. It might also use phase shifting, digital delay, and other such techniques to provide the illusion of a sound stage. The precise means used to achieve this effect and the number of speakers used to do so are user agent-dependent; this property merely identifies the desired end result.

Example(s):

```
H1     { azimuth: 30deg }
TD.a   { azimuth: far-right }          /*  60deg */
#12    { azimuth: behind far-right }   /* 120deg */
P.comment { azimuth: behind }          /* 180deg */
```

If spatial-azimuth is specified and the output device cannot produce sounds behind the listening position, user agents should convert values in the rearwards hemisphere to forwards hemisphere values. One method is as follows:

- if 90deg < x <= 180deg then x := 180deg - x
- if 180deg < x <= 270deg then x := 540deg - x

'elevation'

Value:	<angle>	below	level	above	higher	lower	inherit
Initial:	level
Applies to:	all elements
Inherited:	yes
Percentages:	N/A
Media:	aural

Values of this property have the following meanings.

<angle>   Specifies the elevation as an angle, between '-90deg' and '90deg'. '0deg' means on the forward horizon, which loosely means level with the listener. '90deg' means directly overhead and '-90deg' means directly below.

below   Same as '-90deg'.

level   Same as '0deg'.

above   Same as '90deg'.

higher   Adds 10 degrees to the current elevation.

lower   Subtracts 10 degrees from the current elevation.

The precise means used to achieve this effect and the number of speakers used to do so are undefined. This property merely identifies the desired end result.

Example(s):

```
H1    { elevation: above }
TR.a { elevation: 60deg }
TR.b { elevation: 30deg }
TR.c { elevation: level }
```

19.8 Voice characteristic properties

'speech-rate'

Value:	<number>	x-slow	slow	medium	fast	x-fast	faster	slower	inherit
Initial:	medium
Applies to:	all elements
Inherited:	yes
Percentages:	N/A
Media:	aural

This property specifies the speaking rate. Note that both absolute and relative keyword values are allowed (compare with 'font-size'). Values have the following meanings.

<number> Specifies the speaking rate in words per minute, a quantity that varies somewhat by language but is nevertheless widely supported by speech synthesizers.

x-slow   Same as 80 words per minute.

slow   Same as 120 words per minute

medium   Same as 180 - 200 words per minute.

fast   Same as 300 words per minute.

x-fast   Same as 500 words per minute.

faster   Adds 40 words per minute to the current speech rate.

slower   Subtracts 40 words per minutes from the current speech rate.

'voice-family'

Value:	[[<specific-voice> | <generic-voice>],]* [<specific-voice> | <generic-voice>] | inherit
Initial:	depends on user agent
Applies to:	all elements
Inherited:	yes
Percentages:	N/A
Media:	aural

The value is a comma-separated, prioritized list of voice family names (compare with 'font-family'). Values have the following meanings.

<generic-voice>   Values are voice families. Possible values are 'male', 'female', and 'child'.

<specific-voice>   Values are specific instances (e.g., comedian, trinoids, carlos, lani).
    Example(s):

```
H1 { voice-family: announcer, male }
P.part.romeo  { voice-family: romeo, male }
P.part.juliet { voice-family: juliet, female }
```

Names of specific voices may be quoted, and indeed must be quoted if any of the words that make up the name does not conform to the syntax rules for identifiers. It is also recommended to quote specific voices with a name consisting of more than one word. If quoting is omitted, any whitespace characters before and after the font name are ignored and any sequence of whitespace characters inside the font name is converted to a single space.

'pitch'

Value:	<frequency>	x-low	low	medium	high	x-high	inherit
Initial:	medium
Applies to:	all elements
Inherited:	yes
Percentages:	N/A
Media:	aural

Specifies the average pitch (a frequency) of the speaking voice. The average pitch of a voice depends on the voice family. For example, the average pitch for a standard male voice is around 120Hz, but for a female voice, it's around 210Hz.

Values have the following meanings.

<frequency> Specifies the average pitch of the speaking voice in hertz (Hz).

x-low, low, medium, high, x-high These values do not map to absolute frequencies since these values depend on the voice family. User agents should map these values to appropriate frequencies based on the voice family and user environment. However, user agents must map these values in order (i.e., 'x-low' is a lower frequency than 'low', etc.).

'pitch-range'

Value:	<number>	inherit
Initial:	50
Applies to:	all elements
Inherited:	yes
Percentages:	N/A
Media:	aural

Specifies variation in average pitch. The perceived pitch of a human voice is determined by the fundamental frequency and typically has a value of 120Hz for a male voice and 210Hz for a female voice. Human languages are spoken with varying inflection and pitch; these variations convey additional meaning and emphasis. Thus, a highly animated voice, i.e., one that is heavily inflected, displays a high pitch range. This property specifies the range over which these variations occur, i.e., how much the fundamental frequency may deviate from the average pitch.

Values have the following meanings.

<number>   A value between '0' and '100'. A pitch range of '0' produces a flat, monotonic voice. A pitch range of 50 produces normal inflection. Pitch ranges greater than 50 produce animated voices.

'stress'

Value:	<number> I inherit
Initial:	50
Applies to:	all elements
Inherited:	yes
Percentages:	N/A
Media:	aural

Specifies the height of "local peaks" in the intonation contour of a voice. For example, English is a stressed language, and different parts of a sentence are assigned primary, secondary, or tertiary stress. The value of 'stress' controls the amount of inflection that results from these stress markers. This property is a companion to the 'pitch-range' property and is provided to allow developers to exploit higher-end auditory displays.

Values have the following meanings.

<number>   A value, between '0' and '100'. The meaning of values depends on the language being spoken. For example, a level of '50' for a standard, English-speaking male voice (average pitch = 122Hz), speaking with normal intonation and emphasis would have a different meaning than '50' for an Italian voice.

'richness'

Value:	<number>	inherit
Initial:	50
Applies to:	all elements
Inherited:	yes
Percentages:	N/A
Media:	aural

Specifies the richness, or brightness, of the speaking voice. A rich voice will "carry" in a large room, a smooth voice will not. (The term "smooth" refers to how the wave form looks when drawn.)

Values have the following meanings.

<number> A value between '0' and '100'. The higher the value, the more the voice will carry. A lower value will produce a soft, mellifluous voice.

19.9 Speech properties

An additional speech property, speak-header, is described in the chapter on tables

'speak-punctuation'

Value:	code	none	inherit
Initial:	none
Applies to:	all elements
Inherited:	yes
Percentages:	N/A
Media:	aural

This property specifies how punctuation is spoken. Values have the following meanings.

code Punctuation such as semicolons, braces, and so on are to be spoken literally.

none Punctuation is not to be spoken, but instead rendered naturally as various pauses.

'speak-numeral'

Value:	digits | continuous | inherit
Initial:	continuous
Applies to:	all elements
Inherited:	yes
Percentages:	N/A
Media:	aural

This property controls how numerals are spoken. Values have the following meanings.

digits Speak the numeral as individual digits. Thus, "237" is spoken "Two Three Seven".

continuous Speak the numeral as a full number. Thus, "237" is spoken "Two hundred thirty seven". Word representations are language-dependent.

en Informatique et en Automatique, Keio University). All Rights Reserved. `http://www.w3.org/Consortium/Legal/`" (Hypertext is preferred, but a textual representation is permitted.)

3. *If it exists*, the STATUS of the W3C document.

When space permits, inclusion of the full text of this **NOTICE** should be provided. We request that authorship attribution be provided in any software, documents, or other items or products that you create pursuant to the implementation of the contents of this document, or any portion thereof.

No right to create modifications or derivatives of W3C documents is granted pursuant to this license.

THIS DOCUMENT IS PROVIDED "AS IS," AND COPYRIGHT HOLDERS MAKE NO REPRESENTATIONS OR WARRANTIES, EXPRESS OR IMPLIED, INCLUDING, BUT NOT LIMITED TO, WARRANTIES OF MERCHANTABILITY, FITNESS FOR A PARTICULAR PURPOSE, NON-INFRINGEMENT, OR TITLE; THAT THE CONTENTS OF THE DOCUMENT ARE SUITABLE FOR ANY PURPOSE; NOR THAT THE IMPLEMENTATION OF SUCH CONTENTS WILL NOT INFRINGE ANY THIRD PARTY PATENTS, COPYRIGHTS, TRADEMARKS OR OTHER RIGHTS.

COPYRIGHT HOLDERS WILL NOT BE LIABLE FOR ANY DIRECT, INDIRECT, SPECIAL OR CONSEQUENTIAL DAMAGES ARISING OUT OF ANY USE OF THE DOCUMENT OR THE PERFORMANCE OR IMPLEMENTATION OF THE CONTENTS THEREOF.

The name and trademarks of copyright holders may NOT be used in advertising or publicity pertaining to this document or its contents without specific, written prior permission. Title to copyright in this document will at all times remain with copyright holders.

Glossary

accessible   Information, regardless of form, structure or presentation, that can be easily accessed by any person, regardless of ability.

ADA   See *Americans with Disabilities Act.*

adaptive technology   Hardware and/or software created or modified to enable people to use an interface with or without its standard input or output devices.

administrator   The individual who manages a Web site. See *Webmaster.*

ALS   Amyotrophic Lateral Sclerosis. ALS is a fatal, neuromuscular disease that causes rapid deterioration of motor cells in the brain and spinal cord, ultimately leading to severe impairment of mobility, speech, and respiratory functions. It is more commonly known as Lou Gehrig's disease, after the great New York Yankees baseball player.

Americans with Disabilities Act   U.S. public law enacted in 1990 guaranteeing rights for people with disabilities. This law mandates reasonable accommodation and effective communication, important factors in the justification of a Web site's level of accessibility.

applet   A Java program that is executed within a Web page.

assistive technology Any item, equipment, or product that is used to increase, maintain, or improve functional abilities of individuals with disabilities.

Assistive Technology Act The Assistive Technology Act provides federal aid to states for the development of programs that assist people with disabilities in the purchase of assistive technology devices and services.

authoring tool A software application used to create Web pages and Web sites.

awareness The process of disseminating important information through education and outreach programs. The primary purpose of this book is to build awareness about Web accessibility for people with disabilities.

blindness Blindness includes a variety of conditions involving extreme (but not complete) or complete loss of vision.

Bobby A Web page accessibility validation service located at `http://www.cast.org/bobby`.

Braille A system of writing for individuals who have visual disabilities. The Braille system includes letters, numbers, and punctuation made up of raised dot patterns.

browser The software you use to access and navigate the Web.

Cascading Style Sheets Style sheets can be used to specify color, font types and sizes, spacing, and aural cues on your Web page. Cascading Style Sheets enable you to specify more than one level of style sheet.

Casey Martin Syndrome A term I use to describe the legal quandary industry could face if corporate Web sites are not accessible to people with disabilities. In short, employees with disabilities whose job requires access to a Web site that contains inaccessible content would have grounds for a lawsuit against their employer based on the Americans with Disabilities Act.

client The application used by a person to render or view Web information. See *browser*. See also *user agent*.

CML   Acronym for Chemical Markup Language, a language developed by Peter Murray-Rust.

cognitive disability   A disability involving a person's capacity for processing information and knowledge.

color blindness   The inability to distinguish between combinations and/or pairs of colors.

content developer   An individual responsible for creating information stored on a Web site. Sometimes this person is also responsible for the design and presentation of that information.

CSS   Acronym for Cascading Style Sheets. CSS1 and CSS2 are the W3C official recommendations for style sheets.

deafness   The condition of a person who cannot hear at all.

DECtalk   A speech synthesizer for computers. DECtalk is available as a hardware and software option.

disability   According to the Americans with Disabilities Act (ADA), "The term disability means, with respect to an individual (A) a physical or mental impairment that substantially limits one or more of the major life activities of such individual; (B) a record of such an impairment; or (C) being regarded as having such an impairment."

document type definition   A formal set of markup declarations which specify the rules that an HTML document's markup must follow.

DTD   See *document type definition*.

element   According to ISO 8879, an element is "A component of the hierarchical structure defined by a document type definition; it is identified in a document instance by descriptive markup, usually a start-tag and end-tag."

end tag   Markup that identifies the end of an element.

Fifth Principle   The fifth of seven principles defined by the U.S. National Information Infrastructure Advisory Council (NIIAC). This principle states that "individuals with disabilities should have access to the NII. . . ."

freeware   Computer software that is provided to the general public at no cost.

GII   See *Global Information Infrastructure.*

Global Information Infrastructure   The converging international infrastructure of high-speed, interactive networks transmitted through satellite, terrestrial, and wireless technologies that deliver information and content, regardless of form.

hearing disability   The condition of a person who experiences partial or total loss of hearing.

HTML   Acronym for HyperText Markup Language, the primary language used to create Web pages.

human-centered design   User interface design that focuses on the needs, preferences, and requirements of the user, resulting in a product or process that is accessible and usable.

image   A graphic element stored on a Web page.

image map   A graphic element stored on a Web page that contains regions that are hyperlinks.

Java   An object-oriented programming language invented by James Gosling of Sun Microsystems. The Java platform contains significant features that make it easier for programmers to create Web applications that are accessible people with disabilities.

Java Accessibility API   The Java protocols designed to give assistive technologies access to information in user interface objects.

JavaScript   An object-oriented scripting language developed by Netscape Communications Corporation.

low vision The term used to describe a person who is without a significant measure of vision but is not totally blind.

markup A textual description of data that provides information about the
data type.

MathML A markup language used to describe mathematical equations
and expressions.

Microsoft Active Accessibility Microsoft Active Accessibility (MSAA)
consists of a series of underlying operating system components that improve
the way programs and accessibility aids work together by exposing as much
information as possible about the Microsoft operating systems constructs
(toolbars, menus, text, graphics, and so on) to the accessibility application
or device, as well as to the user.

MSAA See *Microsoft Active Accessibility.*

National Information Infrastructure A nation's internal infrastructure
of high-speed, interactive networks transmitted through satellite, terrestrial
and wireless technologies that deliver information and content, regardless of
form.

navigate The ability for any person to move around within a Web site.

navigation bar Usually a graphical entity that appears on a Web page to
provide navigational information to the user.

NII See *National Information Infrastructure.*

pervasive accessibility The ability of an interface to adapt to the user,
anytime and anywhere.

physical disability The condition of a person who experiences partial or
total loss of physical ability.

presentation The way you render your Web page and its contents
through a user agent. Presentation includes descriptions of font style or size,
as well as colors used on a Web page.

RSI Repetitive strain injury.

SAMI Acronym for Synchronized Accessible Media Interchange. SAMI is a Microsoft file format specification that enables you to create a file with captioned information. Captioning enables a person with a hearing disability to view a textual rendition of the audio associated with a multimedia event.

screen magnification software A software application that increases the size of text and graphics on a computer screen, making it easier to view.

screen reader software A software application that renders electronic information using a synthetic voice.

Section 255 Section 255 of the Telecommunications Act establishes federal guidelines for access to telecommunication services, equipment, and customer premises equipment.

Section 508 Section 508 of the Rehabilitation Act is U.S. legislation that defines the processes used by the federal government to procure electronic and information technology.

SGML See *Standard Generalized Markup Language*.

SMIL Acronym for Synchronized Multimedia Integration Language. SMIL 1.0 is the W3C recommendation for defining a markup language that describes Web multimedia.

speaking disability The condition a person experiences when he or she is not able to speak.

Standard Generalized Markup Language According to ISO 8879, that Standard Generalized Markup Language is "A language for document representation that formalizes markup and frees it of system and processing dependencies." SGML was invented by Dr. Charles F. Goldfarb.

start tag Markup that identifies the start of an element.

structure The information components within an HTML document. For example: headings, lists, and paragraphs.

Style   The format of an element.

style sheets   Used to define the style of elements on a Web page.

tabular information   Information contained within a table.

tag   A delimiter for a Web page element.

Telecommunications Act   A U.S. Act that includes guidelines and specifications involving access by people with disabilities. As defined by the 104th Congress of the United States, the Telecommunications Act of 1996 is designed to "promote competition and reduce regulation in order to secure lower prices and higher quality services for American telecommunications consumers and encourage the rapid deployment of new telecommunications technologies."

Third Age   The generation of elderly individuals.

Tim Berners-Lee   The guy who invented the World Wide Web.

URL   Acronym for Uniform Resource Locater.

usability   A term used to describe the processes for developing interfaces that are easy to use by people.

user agent   The software you use to access and navigate the Web.

user interface   Any part of a system with which a user interacts.

valid HTML   A Web page or HTML document is considered valid when it complies with W3C HTML recommendations. The HTML 4.0 Strict recommendation includes accessibility constructs.

validation service   An online software service that validates a Web page (or site) according to W3C HTML or W3C WAI recommendations.

visual disability   The condition of a person who experiences partial or total loss of vision.

W3C   Acronym for the World Wide Web Consortium.

WAI   The acronym created by the author for the Web Accessibility Initiative.

Web Content Accessibility Guidelines   The official set of guidelines published by the W3C to assist Web content developers in the creation of accessible Web sites.

Web Accessibility Initiative   The international initiative established by W3C and the Yuri Rubinsky Insight Foundation in 1997. The WAI and its program office works with organizations around the world to promote Web accessibility in five key areas: technology, guidelines, tools, education and outreach, and research and development.

WebABLE!   What the Web should be to all people. Also, the author's Web site located at http://www.webable.com.

Webmaster   The person responsible for a Web site. See *administrator*.

Yuri Rubinsky   The deceased Founder and President of SoftQuad. Yuri and Jeff Suttor were instrumental in integrating accessibility into HTML 2.0. This book is dedicated to Yuri's memory.

Yuri Rubinsky Insight Foundation   The organization created in Yuri Rubinsky's memory. The YRIF, working in conjunction with the W3C, launched the Web Accessibility Initiative in 1997.

XML   Acronym for the eXtensible Markup Language.

Index

A

ABBR attribute 114
ABBR element 101
Able Channel 308
ACB 293
Access Adobe 255
Access Pack 238–239
access technologies 70
AccessDOS 238–239
accessibility
    active 234
    Australian policies and standards 40
    Canadian policies and standards 42
    design issues 88
    HTML 95
    PC 265
    pervasive 272, 377
    Portuguese policies and standards 43
    test 121
    tools 230–231
    U.S. policies and standards 35
    United Kingdom policies and standards 44
    Web 3
    Web guidelines 320
    Web site review 120, 123
accessibility browsers 140

    BrookesTalk 147
    Enhancing Internet Access (EIA) 152
    Home Page Reader 145
    pwWebSpeak Plus 141–143
    Sensus 148–149
    Sigtuna 150
    VIP 151
    WebCite 154
Accessible 215
accessible 336, 373
Accessible Action 215
accessible authoring tools, guidelines 210
accessible design 88, 316
accessible Java programs, guidelines 218
AccessibleBundle 216
AccessibleComponent 215
AccessibleContext 215
AccessibleHyperText 215
AccessibleResourceBundle 216
AccessibleRole 216
AccessibleSelection 215
AccessibleState 216
AccessibleStateSet 216
AccessibleText 215
AccessibleValue 215
ACESSKEY attribute 103

381

X Y Z

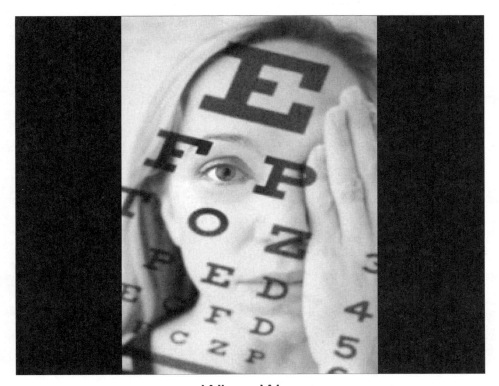

When Was
The Last Time
You Had Your Site Checked?

Everyday your site is tested by people with disabilities.

You should see what they see.

The question is how well are you communicating your message to them? Don't wait for consumer complaints or legal action. It's time to be proactive to ensure your site is accessible to all people. WebABLE reviews your web site and measures design, functionality and accessibility. WebABLE ensures that your web site meets the requirements for accessibility.

Don't Wait — visit www.webable.com or email webinfo@webable.com today!
Learn how WebABLE can make your web site accessible.

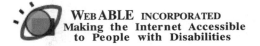

W3C Quick Tips to Make Accessible Web Sites

Web Accessibility Initiative

Images & animations	Use the **alt** attribute to describe the function of all visuals.
Image maps	Use client-side **MAP** and text for hotspots.
Multimedia	Provide captioning and transcripts of audio, and descriptions of video.
Hypertext links	Use text that makes sense when read out of context. For instance, avoid "click here."
Page organization	Use headings, lists, and consistent structure. Use **CSS** for layout and style where possible.
Graphs & charts	Summarize or use the **longdesc** attribute.
Scripts, applets, & plug-ins	Provide alternative content in case active features are inaccessible or unsupported.
Frames	Use **NOFRAMES** and meaningful titles.
Tables	Make line by line reading sensible. Summarize.
Check your work	Validate. Use tools, checklists, and guidelines at `www.w3.org/TR/WAI-WEBCONTENT`

For complete Guidelines and Checklist, see www.w3.org/WAI